Counter-Space Defence Co-Orbital Satellite Fighter

The Soviet Istrebitel Sputnik Anti-Satellite Complex

HUGH HARKINS

Counter-Space Defence
Co-Orbital Satellite Fighter

The Soviet Istrebitel Sputnik Anti-Satellite Complex

Centurion Publishing
United Kingdom

ISBN 10: 1903630673
ISBN 13: 978-1903630679

This volume first published in 2017

The publisher and author would like to thank all organisations and services for their assistance and contributions in the preparation of this volume: Open Joint Stock Company 'Corporation' Space System Special, Comet; Yuzhnoye State Design Office; PJSC Turaevo Machine Building Design Bureau, Soyuz; S.P. Korolev Rocket and Space Corporation, Energia; JSC MIC Mashinostroyenia (Joint Stock Company Military Industrial Corporation Scientific and Production Machine Building Association); JSC NPO Energomash; OSC KBKhA (Konstruktorskoe Buro Khimavtomatiky, Krunichev State Research and Production Space Centre); National Space Agency of Ukraine; State Space Corporation ROSCOSMOS; National Aeronautics and Space Administration, NASA; United States Department of Defence; Central Intelligence Agency and the Ministry of Defense of the Russian Federation

CONTENTS

INTRODUCTION

The Soviet Counter-Space Defence Satellite Fighter complex was the only dedicated anti-satellite system to be operationally deployed during the Cold War. This system, which was allocated to alert duties in 1972 and attained full operational capability in 1978, served beyond the break-up of the Soviet Union in December 1991, being decommissioned by the Russian Federation in 1993.

This volume sets out to detail, from the historical and technological perspectives, the Counter-Space Defence Satellite Fighter complex. The road, factual and propaganda, that led to the development of the ASAT complex is detailed, as well as the Counter-Space Defence Satellite Fighter complex itself, along with the flight test and development program leading to actual deployment of the system.

At varying times throughout the text the various weapon systems discussed will be referred to under their Soviet service and or manufacturer names and designations as well as, at appropriate times, under their NATO shadow designations, the latter of course being accepted by the Soviet Union for use in arms limitations and other treaties.

All technical data concerning the respective weapon systems and their components have been provided by the respective design bureau/offices, as has much of the imagery and graphics with additional impute from United States intelligence agencies and defense department, the space agencies of the Ukraine and the Russian Federation, United States National Aeronautics and Space Administration and the Ministry of Defense of the Russian Federation.

1

IDENTIFYING A PERCEIVED SPACE BASED THREAT

'The Russians can kill us in space' – this was the bleak February 1978 assessment of the United States CIA (Central Intelligence Agency) chief on the survivability of much of the American dominated NATO (North Atlantic Treaty Organisation) space based communications and observation systems in the event of a war with the Soviet Union (USSR – Union of Soviet Socialist Republics). The basis for the somewhat somber tone of the assessment was the distinct lack of any NATO ASAT (Anti Satellite) capability that was not mirrored by the Soviet Union, which possessed a undetermined size force of IS (Istrebitel Sputnik) Co-Orbital Satellite Fighter (Истребитель спутников) spacecraft that could intercept target satellites out to orbits in excess of 1500 km altitude – the realm of the orbits of much of the NATO intelligence/surveillance/communications satellite capability.

As a very young boy growing up in the late 1970's there was no notion of fighter/interceptors in space other than, perhaps, the fabled Rebel 'X-wing' interceptors and Empire 'Tie' fighters in the motion picture 'Star Wars Episode IV' (released in the UK in 1977). While these props of the 'big screen' did not seem too farfetched to the vivid imagination that comes at so young an age, there was no accurate public knowledge of such weapon systems as the Soviet Co-Orbital Space Defence Satellite Fighter that had entered limited operational service almost half a decade before. Therefore, in those pre-high school years, it seemed clear that interceptors in space were a thing of the 'silver screen' and, perhaps, the future, however near or distant. However, on looking back to yesteryear with the benefit of hindsight it seems logical, given the thinking of the time, that the militarisation of space by one or both of the power blocks of East and West was a possibility if not a probability. After all, this was the period in human history when we, as a species, were MAD, a fitting acronym for Mutually Assured Destruction, the policy that ensured that both power blocks would be destroyed in the event of a full-scale nuclear exchange – true Armageddon, although this would have been no 'Old Testament' last battle between good and evil, but the tragic conclusion to a sibling rivalry between evil and her near identical twin, nefarious.

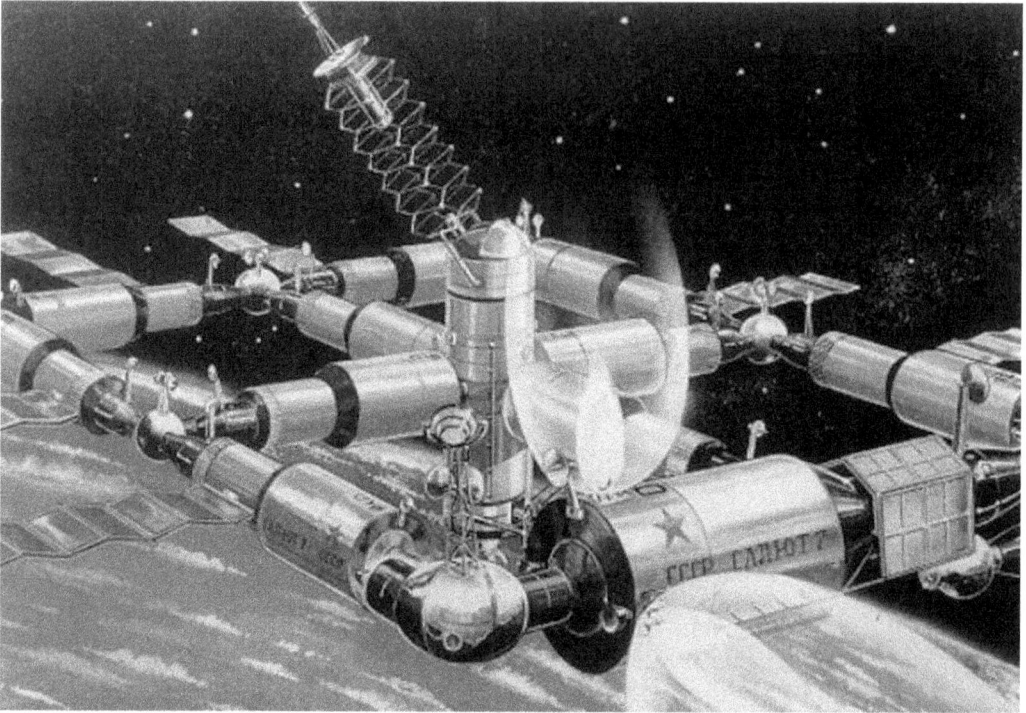

Resembling something from the set of a James Bond movie, this was an artist depiction of an early 1980's assessment of a future Soviet orbiting space station complete with a laser defence system that could threaten other nation's spacecraft. While no such armed orbital stations were planned, the Soviets were, at that time, operating small manned space stations of the Salyut type, which in-turn led to the Mir space station launched from 1986, a more practical, in technological terms, anti-satellite capability did exist in the form of the Co-Orbital Satellite Fighter. DIA

While this volume is not intended to serve as a history of Soviet rocket development, a short overview of such development milestones leading to the launch vehicles that would form major elements of the Counter Space Defence Satellite Fighter complex would seem pertinent. It is an indisputable historical fact that the post-World War II rocket programs in the Soviet Union and the United States that would lead to the dawn of the ICBM (Intercontinental Ballistic Missile) and space payload launch vehicles were born out of the ashes of World War II Germany and in particular the A-4 (V-2, FAU-2)) rocket, which was tested and further adapted by the Soviets and Americans. In the Soviet Union this lead to a 14 April 1948 government decree authorising development of what would become the R-1, the first Soviet ballistic missile derived from the A-4.

The development of the R-1 was enabled by the formation of several test facilities and manufacturing centers, as had been decreed by the Soviet government on 13 May 1946, as part of the overall effort to develop a LRBM (Long Range Ballistic Missile) under the direction of Chief Designer S.P. Korolev. A decree of the Soviet minister of armaments, D.F. Ustinov, issued on 26 August 1946, authorised the formation of an organisational structure under NII-88, effectively putting in place

Department 3 of the Special Design Bureau – OKB-1 (now the S.P. Korolev Rocket and Space Corporation Energia) under the leadership of Korolev who would go on to become the architect of the Soviet Union's early long range ballistic missile and space payload launch vehicle programs that heralded the dawn of the space age.

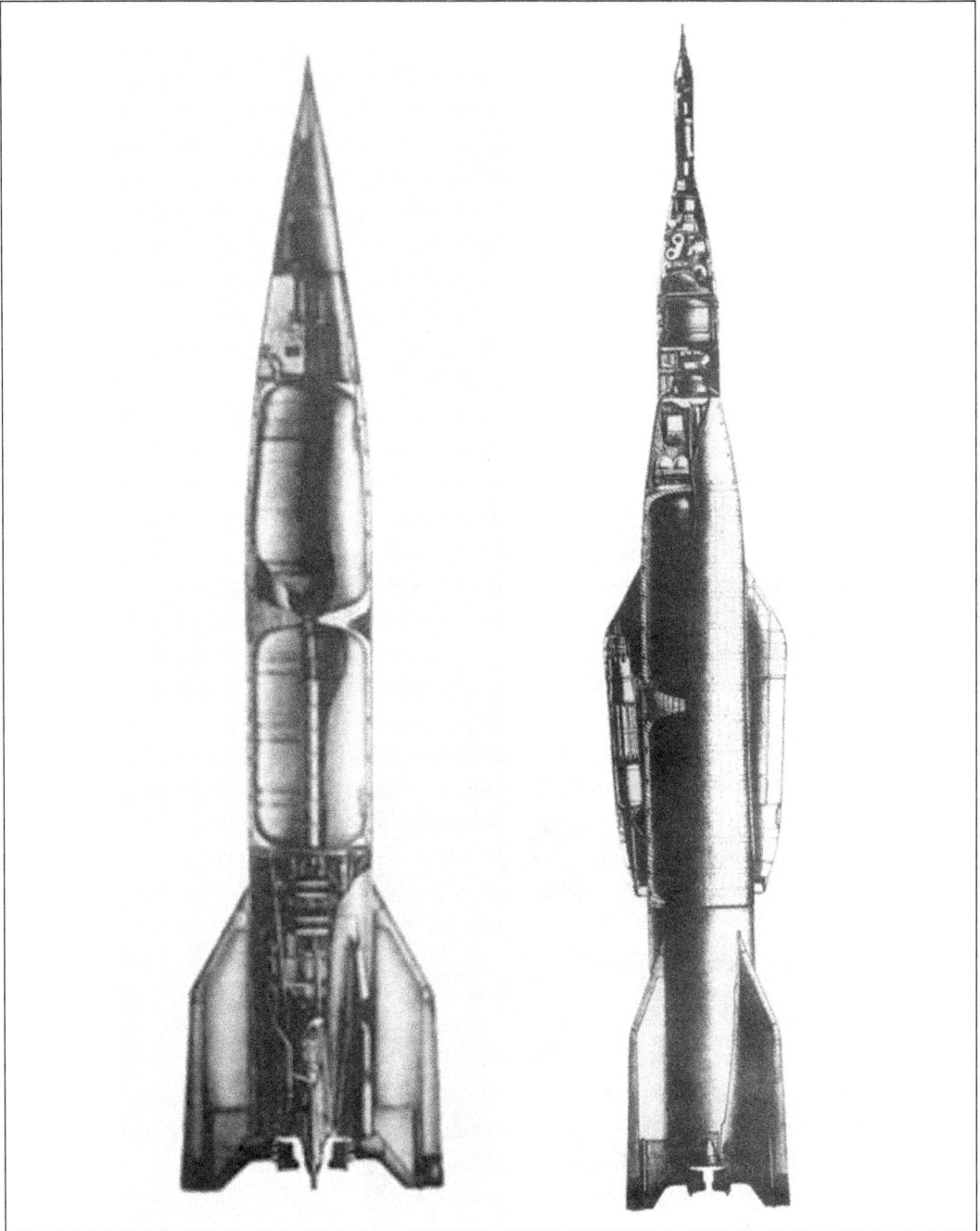

R-1A Ballistic missile (left) and R-1D Geophysical research rocket (right). Note: Both rockets are to different scales. JSC Energia

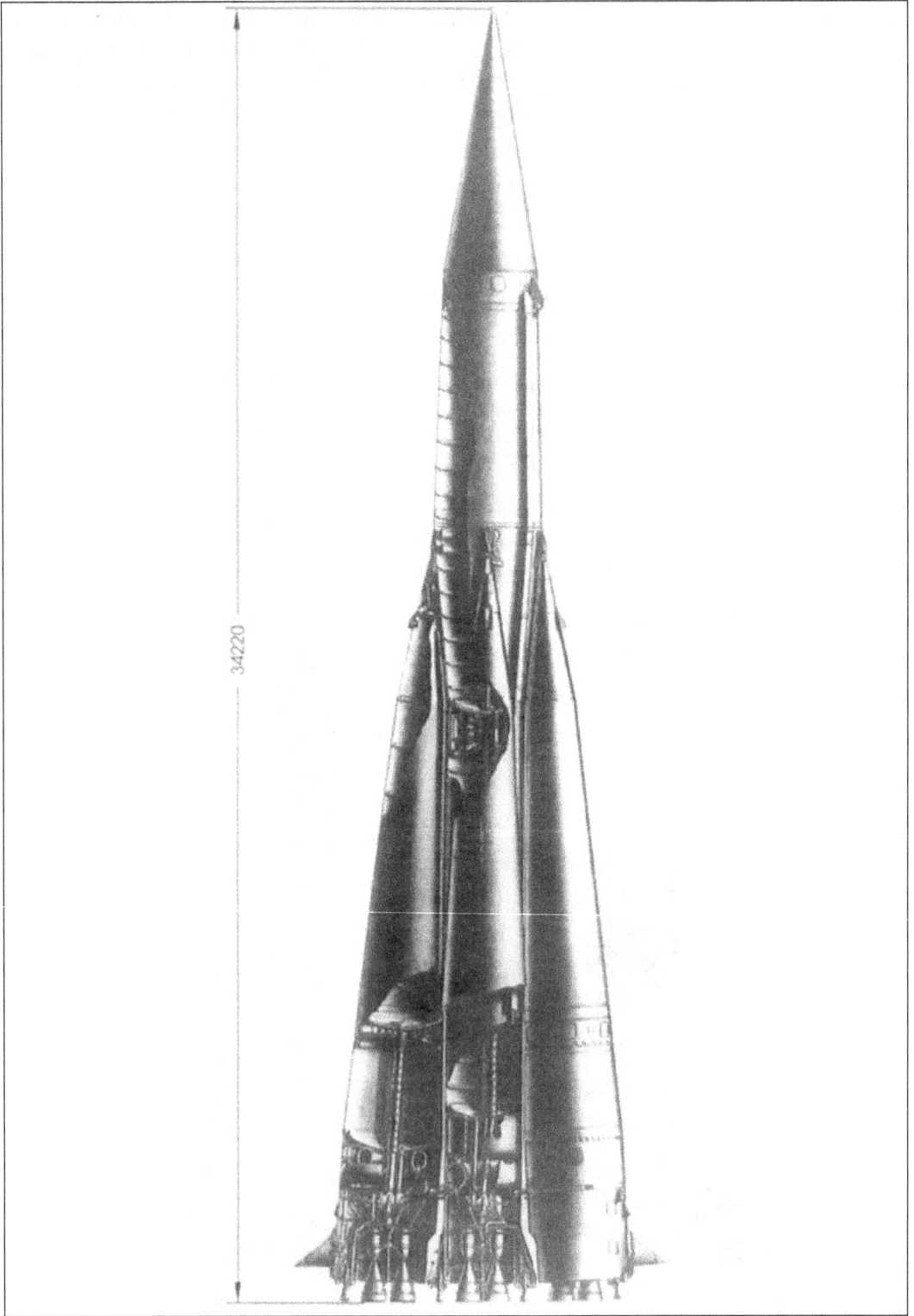

The dawn of the age of space travel came with the launch of Sputnik 1 on a modified 8K71 R-7 ICBM on 4 October 1957. JSC Energia

On 18 October 1947, the first Soviet launch of an A-4 rocket, assembled from various units of captured German rockets, was conducted. Later that year the Soviets concluded A-4 testing, which had paved the way for the first launch of an R-1 rocket in 17 September 1948. In 1950, the R-1 entered Soviet service with a maximum firing range of 270 km. This would bring it under the later SRBM (Short-Range Ballistic Missiles) category of missiles that covered vehicles with a maximum firing range up to 1109 km (599 nm).

Further missile developments led to the R-1A, the first of which was launched on 21 April 1949, the first of a series of six upper atmosphere missions (altitudes of 100 km). The R-2E (Experimental) ballistic missile, the first launch of which was conducted on 21 September 1949, featured a detachable nose-cone for the warhead. The first Soviet special (nuclear) armed ballistic missile complex was the R-5, the first launch of which took place on 15 March 1953. The R-5 was further developed into the R-5M complex, the maiden launch of which took place on 21 January 1955. This missile complex could deliver a special payload to a distance of 1200 km, effectively bringing the weapon into the MRBM (Medium Range Ballistic Missile) category, which covered such missiles with the firing range coverage of 1111 km (600 nm) to 2778 km (1500 nm).

Although R-5 variants had been instrumental in bestowing upon the Soviet Union a nuclear armed ballistic missile capability, it was with the R-7 (NATO reporting designation/and name SS-6 'Sapwood') that the dawn of the ICBM and space travel would become a reality. This missile complex, which had been developed under a Soviet government decree issued on 13 February 1953, at a stroke, temporarily, catapulted the Soviet Union ahead of its competitors in both fields. The two-stage R-7 incorporated a fire charge mass of up to 3000 kg, with a total warhead mass of up to 5500 kg, and was capable of striking targets 8000 km distant from the launch site, bringing it well beyond the 5556 km milestone that would define the ICBM category of missiles.

The R-7, which entered service as an ICBM in August 1957, would lead to a whole family of space payload launch vehicles that would include the Vostok and Molniya, which would lead to the Soyuz launch vehicle. The first ever payload delivery to Earth orbit was conducted on 4 November 1957, when a modified R-7 ICBM launched the world's first artificial satellite, Sputnik 1 (PS-1), setting the Soviet Union apart from the rest of the field in regards to development of space launch vehicles, in effect the stone that started the ripple of shock waves that resonated around the world in the form of the *bleep, bleep, bleep…* transmitted from the satellite back to Earth. The R-7 launch vehicle derivatives would be instrumental in the Soviet Union's capacity to maintain a lead in unmanned and manned spaceflight for several years. Such launch vehicles proved to be suitable for near Earth orbit uninhabited and inhabited space flights, as well as several types of uninhabited Moon missions and interplanetary probes to Venus and Mars. Indeed, it was an R-7 derivative that would launch the first hardware directly associated with the Soviet ASAT program, namely the Polyet-1 and Ployet-2 manoeuvrable satellite demonstrator spacecraft in 1963 and 1964 respectively.

While OKB-1 had certainly been the most high profile of the Soviet ballistic missile designers, other design bureau were heavily involved in the development of these promising strategic strike platforms. In this regard, NII-88 forwarded to Vasily Budnik, Chief Designer at Dnepropetrovsk Factory No.586, plans and research data on what was referred to as a 'high boiling' missile on a par with the OKB-1 R-5. This would lead to the development and building of the prototype of the R-12 long-range ballistic missile under OKB-586, also referred to as SDB-586, as Factory No.586 had been renamed in 1954 (now the Ukrainian Yuzhnoye State Design Office), headed by its chief designer, Mikhail Yangel. The R-12, the first launch of which was conducted on 22 June 1957, had a maximum firing range of 2080 km, bringing it into the MRBM category.

A modern day computer generated graphic depicting the Sputnik-1 artificial Earth satellite orbiting the planet in November 1957. JSC Energia

Continued research and development led to the R-14, which was endowed with improved accuracy and a maximum firing range of 4500 km, more than twice that of the R-12, bringing this missile into the IRBM category covered by ranges from 2778 km up to 5556 km. Further research and development led to the R-16, which, with a maximum firing range of 13000 km, was OKB-586's first ICBM complex to reach production. The first launch was prepared for under an atmosphere of accelerated pace. A faulty control system cable network was the catalyst for a premature start of the missile second stage sustainer resulting in the missile being destroyed in a catastrophic explosion on the launch pad. Following redesign to eliminate the faults an R-16 missile was successfully launched in February 1961.

While the R-12, R-14 and R-16 constituted the first generation of ballistic missiles developed by OKB-586, further missile development by the bureau would lead to the second generation R-36 ICBM and the world's first deployable orbital/fractional orbit weapon system that formed the basis for the Cyclone-2 space payload launch vehicle that would form a major element of the Counter Space Defence complex.

The Soviet ASAT capability, from which the first ever interception of a spacecraft by another spacecraft was conducted on 1 November 1968 when Cosmos 252 intercepted Cosmos 248, was developed from the late 1950's through the late 1970's. This had commenced at a time of Soviet nuclear inferiority in comparison to NATO. By the time of the first successful ASAT tests in late 1968, the Soviets were well on the way to achieving nuclear parity with NATO, which was achieved by the time the ASAT system had entered limited service in 1972; Soviet nuclear superiority being achieved before the ASAT system entered full operational service in 1978.

Of course, the perceived need for an ASAT system and, indeed, the catalyst that would allow the development of such a system, was born out of the same event – the launch of the world's first artificial Earth satellite, Sputnik-1 on 4 October 1957. This relatively innocuous start to human space flight would soon lead to the realisation of long theorised uses of space, for those nations capable of developing a space launch capability, against ones enemies, actual and perceived. Long before the dawn of the space age, brought in by that '*bleep, bleep, bleep*' of the relatively simple transmitter on-board Sputnik-1, there had been much planning for the potential use of Earth orbit for the stationing of offensive weapons. Only a few such programs were considered viable and non-carried through to fruition save the Soviet FOBS (Fractional Orbit Bombardment System), which would remain terrestrial based until required for use.

Even prior to the dawn of the space age that commenced that fateful October day in 1957, there was detailed discussion of the use 'outer space' as was then the term for what lay beyond the Earth's atmosphere, to base orbiting platforms capable of launching atomic bombs on targets back on Earth. Among the most prominent of such early theories to come out of the Western block were those touted by the former German scientist Dornberger, who, with the end of World War II in Europe in summer 1945, was transferred to the United States to work on American funded rocket programs. This particular scientist was scoffed at as a former Hitlerite by the Soviet Union, which held the view that he was nothing less than a deranged madman who, now under his new masters in the United States, continued to envision the same destruction of Bolshevism advocated by the former German Chancellor, Adolf Hitler, universally considered responsible for the deaths of tens of millions of people, during the terrible war years collectively referred to as World War II, but known as the Great Patriotic War in the Soviet Union.

The fact that Hitler's victims predominantly came from Eastern Europe and the USSR was not lost on a Soviet Union perplexed at such statements coming from the likes of Dornberger who, it appeared clear by virtue of those same statements, was still envisioning the destruction of the Soviet Union. The United States intelligence document on 'Soviet Orbital Rockets', 62-104279, stated that as early as 1948, almost a decade before the first successful launch of an artificial satellite, Dornberger had proposed that, in the event of developments allowing access to space, "the earth be surrounded by hundreds of artificial satellites in the form of nuclear bombs." The document then went on to directly quote, supposedly Soviet writings of a Dornberger quote, that "the orbits of these bombs must be laid over Russia. I see no reason to prevent us [inferred to be the United States] from doing so… In case of war we will not have to change orbits. I think that it would be possible to launch these bombs with sufficient precision." Perhaps most troubling from this text, from a Soviet point of view, was that in 1948 the Soviet Union was not yet a nuclear power (the first Soviet atomic device was detonated in 1949) and here was what appeared to be detailed plans for the destruction of the country's infrastructure and population by nuclear weapons in the event of the outbreak a conventional war.

It was such visions of the use of space as a platform for the launching of atomic weapons to destroy the cities of the Soviet Union that spurred Soviet attitudes towards achieving superiority in space in both the fields of exploration and, if need be, for military advantage. For this, the Soviets would capitalise on their early superiority in rocket development, evinced by the launch of Sputnik 1, which was followed by what seemed to be Soviet success after Soviet success in the areas of uninhabited and inhabited spaceflight over the next half decade or so, encompassing a period known as the early space age.

During the period of the early space age there were many idle references in the Soviet and western press about the feasibility of space based weapons, such references to satellites that could launch hydrogen bombs against targets on Earth continuing through the early 1960's, although there was no official references to such theorised weapon systems. Soviet research efforts were being invigorated by intelligence evidence coming out of the United States on such space based weapons programs as Project 'Bambi', Project 'Saint' and Project 'Sped', not to mention the potential usage of the X-20 Dynasoar spaceplane, then under development, as what was being unofficially referred to as a potential orbital bomber.

Concern about the possible deployment of orbital weapons by the United States led to Soviet overtures that proposed a ban on the deployment of offensive weapons in space. In reference to possible future weapons applications in space, Professor A. Rybkin, in an article published in the Soviet publication Red Star on 22 March 1958, acknowledged the feasibility of placing "atomic and hydrogen" weapons, presumably bombs, on orbiting satellites. Rybkin's statement went on to read "As the foreign press has stated, satellites can carry atomic and hydrogen weapons. Considering the difficulties in intercepting and destroying a satellite in space, it must be kept in mind that it could be a dangerous weapon in the hands of the aggressor. The Soviet Government's proposal, which poses banning the utilization of space for military purposes, also includes conditions for international control, which would preclude the possibility of artificial earth satellites being used for aggressive purposes."

There were of course many references in the western press in regards to the development of orbital weapon systems of the type that could deploy atomic and hydrogen bombs. The first official references to orbital bombardment systems surfaced at a Soviet Trade Unions Congress speech in Moscow by Soviet Premier Nikita S. Khrushchev on 8 December 1961. Khrushchev dropped passive hints about the feasibility of the Soviet Union developing such systems on the back of recent space exploration exploits.

For all the unofficial and official references to the potential deployment of atomic weapons in space, it was clear, even from the earliest post Sputnik days, that the predominant military use of space would probably come down to surveillance and communications roles. It was in the area of overhead observation that the new space technologies offered most promise and through the 1960's many so-called 'spy' satellites were launched as each of the two major power blocks of East and West (the Warsaw Pact and NATO respectively) kept a close watch on each other's activities.

The Discoverer 1 satellite that was placed into Earth orbit by the United States in 1959 had been regarded by the Soviet Union as a direct threat to her sovereignty, as this was the first real test of the system designed for surveillance. The year 1959 had also seen the commencement by the United States of Project 'Saint', widely deemed to have been an early attempt to produce an ASAT capability, or at the very least a capability to rendezvous with a satellite for the purposes of inspection. This potential threat of future interference was the catalyst for the future implementation of a Soviet ASAT program at the behest of the Soviet Premier, Khrushchev, as the Soviet Union was resolute in its determination that it would not be denied access to and the use of space. These tentative first steps on the path toward an operational Soviet ASAT system was effectively the conception of what would evolve into what is now termed missile and space defence, which would encompass ballistic missile attack early warning, space surveillance/tracking, ballistic missile defence and space systems defence.

In the 1960's, there was no shortage of fanciful ideas on ways to interfere with other nations space operations, some of which, I do confess, dragged out a chuckle or two whilst they were being researched. Such ideas were not always by means of direct attack with a weapon system of some sort or other. One such theorised operation would be to, in some way, force an adversaries spacecraft in a low Earth orbit into a denser part of Earth's atmosphere where its orbit would decay and the spacecraft would burn up as it reentered the atmosphere proper. Among the crazy ideas of how such a feat would be accomplished was laid out in United States intelligence document on 'Soviet Orbital Rockets', 62-104279, which stated "it is proposed to create piloted fighter apparatuses, part of whose crews could temporarily leave their ship, and attach braking rocket engines to it [the target spacecraft]." These however were not Soviet plans, but a Soviet inference on plans the US possibly had through the interpretations, indeed misinterpretations of interviews and writings in the US press.

More credible in terms of operational realities was the potential to develop uninhabited spacecraft that could rendezvous with another spacecraft for the purpose of inspection and or destroying the target spacecraft. As stated by Col. Malishkin in Red Star, 8 August 1963 (republished in Soviet Orbital Rockets', 62-104279), such inspector satellites assumed to be under development under 'Project Saint' would be equipped with "a radar set and computer with memory device... for determining the orbit of a new satellite. The memory device would hold data on the speed and orbital figures on the inspector satellite, as well as those of all known satellites in orbit. By comparing the orbital parameters it would be possible to determine the appearance of a new satellite. If this takes place, the 'Saint' would send a report to earth, approach the satellite and determine its purpose with the aid of equipment on board". The text continued, "In addition to the radar equipment, the inspector satellites will contain television equipment for transmitting an image of the satellite being identified to earth in order to determine whether the satellite carries a nuclear weapon, a radiation indicator would be mounted on the 'space inspector'. Other units will make it possible to determine the mass of the satellite, which will help in drawing conclusions as to its military purpose. The inspector satellite will be put into orbit somewhat higher and ahead of the satellite being identified. Moving at a great rate of speed in orbit, the 'Saint' would 'intercept' the target and then, reducing its velocity with rocket engines, would draw close to it. The initial distance for target interception would be about 80 kilometers. In order to identify a satellite the 'Saint' would have to move to a distance of 15-30 meters from it". The Soviet impression was that an improved variant of the any inspector satellite developed under Project Saint would be used to achieve a hard kill by carrying some kind of weapon system, effectively turning the inspector satellite into an interceptor satellite. However, what such writings as that of Col. Malishkin showed was not potential operational scenarios for 'Project Saint' or, indeed, 'Project Bambi', but an early look at the potential operating parameters for a future Soviet ASAT interceptor, which was, at that time, already under development. It was clear that the Soviets intended to overcome then current technological shortcomings in spacecraft interception potential in an effort to create a counter to any US plans to dominate Earth orbit.

As well as the potential for an ASAT capability that could be launched into Earth orbit to conduct a co-orbital interception for the purposes of inspector or destruction, both power blocks of East and West were also heavily involved in ABM (Anti-Ballistic Missile) defence systems to counter the growing threat from ICBM's. These ABM systems, it was clear, could be developed with an inherent ASAT capability against spacecraft in very low Earth orbit.

US plans for ASAT variants of the Nike-Zeus SAM (Surface to Air Missile) and possibly the Thor ICBM were contingent on these weapons being equipped with powerful nuclear warheads to achieve a kill as the systems were not accurate enough to destroy a spacecraft with a conventional armed warhead. It was widely recognised, within both the major power blocks, that employing nuclear warheads to try to deny an enemy the use of space would also have the net result of denying your own forces the use of space based assets as the resultant explosions could destroy friendly spacecraft as well as enemy spacecraft. Even if beyond the lethal range of the blast, the effects of nuclear explosions in low-Earth orbit would, in all probability, knock out systems or perhaps induce small adjustments in orbit that would render the craft mission ineffective. These drawbacks, among other reasons, were the deciding factors in the abandonment of the Nike-Zeus/Thor missiles as viable ASAT systems.

In the Soviet Union, a first generation ABM system was brought to fruition in the shape of the A-35 (NATO reporting name Galosh') complex positioned in a ring around the Soviet capital, Moscow. This system, which was retired with the introduction of its successor from the mid-1990's, was credited with having a rudimentary ASAT capability although the main ballistic missile warhead engagement scenario for the non-nuclear armed weapon envisioned the target being struck inside the Earths atmosphere. Even if employed as an ASAT system, the A-35, like all potential ground based ABM missile system, would have been restricted in engagement arc and altitude, in this case against satellites orbiting over the central or western Soviet Union in very low orbits only - the realm of perhaps a few phot-reconnaissance satellites. In order to intercept satellites at orbits higher than a few hundred miles it was clear that a co-orbital satellite interceptor would be required.

Artist depiction of a Soviet A-35 Anti-Ballistic Missile being launched. This system, deployed around Moscow, could, it was inferred, be adapted to incorporate a rudimentary ASAT capability against targets in very-low Earth orbit, but only when over a very small part of the spacecraft orbit. DIA

It is clear from Soviet writings that the impetus behind the development of their co-orbital ASAT program was the fear that the US was developing ASAT systems under its 'Project Saint' as well as so called inspector satellites under 'Project Bambi', in an effort to dominate space. It was statements of the need, and even intent, of the US to perform close inspection of Eastern bloc spacecraft, combined with the potential threat that the US may orbit nuclear weapons under any future MOBS (Multiple Orbit Bombardment System) that prompted the Soviet Union to vigorously pursue a viable ASAT interceptor program capable of combating surveillance satellites and any potential MOBS system. The Soviet Union, determined that it would not be excluded from space operations of either civil or a military nature, continued apace with its own Counter-Space Defence ASAT program which would draw on some technology being developed under its planned FOBS development program.

The reasoning behind the Soviet decision to adopt an ASAT capability was in part to protect projected systems such as its manned space stations Salyut type (above) from potential interference by other nations. Ironically, NATO mistakenly associated such space stations as being part of Soviet space defence planning. JSC Energia

2

COUNTER-SPACE DEFENSE CO-ORBITAL SATELLITE FIGHTER COMPLEX

The Soviet IS (Istrebitel Sputnik) Counter-Space Defence Co-Orbital Satellite Fighter Complex was developed as a viable ASAT (Anti-Satellite) system designed, as stated in documentation of TMBDB Soyuz, engine developer for the Satellite Fighter (Истребитель спутников), to "stop active functioning of those space vehicles that are targeted". The counter-space defence complex would comprise a number of major components - the Cyclone-2 (Tsyclone-2) space payload launch vehicle, the Co-Orbital Satellite Fighter spacecraft that would intercept target satellites, the launch base and technical facilities and the command and control centre(s). Each of these main components could be broken down into a larger number of sub-components, not least of which was the target satellites employed during the development of the complex.

Feasibility studies for the Istrebitel Sputnik program were initiated by KB-1, a subsidiary of OKB-41, under a program of what loosely translated into English as 'cosmic shock', information and control and intelligence system. The studies, begun in 1959, led to the establishment of an umbrella organisation for the creation of an IP Satellite Fighter headed by OKB-52 – EDB Experimental Design Bureau 52 (now JSC MIC Mashinostroyenia), under chief designer Vladimir Nikolaevitch. Chelomey, later that year.

The Cyclone-2 space-payload launch vehicle that would constitute the launch vehicle component of the projected IS was a development of the 8K69 FOBS (Fractional Orbit Bombardment System), which was itself developed from the 8K67 ICBM (Intercontinental Ballistic Missile). The 8K69 was itself the end product of Soviet efforts to develop a FOBS that had transitioned from the theoretical to the practical development stage with the Kremlin's decision to develop such a system from 1961. Initially such a weapon system was to be based on an OKB-52 (EDB-52) design based on the UR-200 launch vehicle. Development of the UR-200 had commenced in 1960 as a first generation space launch vehicle/heavy ICBM, design solutions being worked so that the main engine thrust could be deviated in direction.

During the 1961-1964 timeframe, the OCB-154 (OKB-154) (now OSC KBKhA (Konstruktorskoe Buro Khimavtomatiky) RD-203 and RD-204 rocket engines were developed for the first stage of the UR-200, along with the second stage RD-205, which incorporated the RD-206 main engines and RD-207 steering engines. All of these engines, which burned storable propellants, introduced the staged combustion cycle schematic that conferred a double combustion chamber pressure of up to 150 kg/cm^2 compared to the 70 kg/cm^2, which was the norm for open cycle engines.

The UR-200 ICBM was among the early grouping of projects deemed to have the potential for further development as the launch vehicle for the Soviet Unions planned space weapon systems that were in the design/development stage in the early 1960's. Krunichev

The UR-200 possessed characteristics that appeared to facilitate the potential for development as a so called 'global rocket'/FOBS and, if the Kremlin deemed such a move necessary, further development as the launch vehicle component of a potential MOBS (Multiple Orbit Bombardment System), which would, unlike a FOBS, be deployed in Earth orbit in peacetime. Besides the FOBS and MOBS, the UR-200 was intended to constitute the launch vehicle component for other potential space-based weapons systems such as ASAT systems. This was an area in which EDB-52 was involved in development work for a potential deployable ASAT interception capability, to which end the design bureau designed and built the words first maneuvering satellite, the 'Polyet-1', which was launched on 1 November 1963, this program being pursued in conjunction with other development work on ASAT systems.

The Proton-K (UR-500K) launch vehicle was developed from the UR-500 launch vehicle which was initially designed to form the basis of an orbital weapon complex launch vehicle following the demise of the UR-200 and the OKB-1 GR-1. Krunichev

The first launch of a UR-200, which had a launch weight in the order of 140 tons, was conducted on 4 November 1963, less than two months after the first launch of an 8K67 ICBM of the OKB-586 (Yuzhnoye State Design Office in Soviet Ukraine (now independent)) second-generation R-36 missile complex. This was followed by a further eight successful launches that would validate the successful operation of the rocket system, which, however, fell short on the operational requirements specified for the system. The major shortfall was that it was not able to stand for extended time periods after it had been fueled. This, together with other pre-launch preparation requirements, including the need to be transported from the storage facility and sealing, added up to a system very vulnerable to a NATO (North Atlantic Treaty Organisation) first strike. The shortfalls in capability resulted in the programs termination as an ICBM/military space payload launch vehicle and, ultimately as a potential platform for any future orbital/fractional orbit or ASAT weapon complex.

While the UR-200 program was still moving forward, OKB-1 (now S.P. Korolev Rocket and Space Corporation, Energia) was in the early design stages of its GR-1 program, which was officially authorised by a decree of the Soviet government dated 24 September 1962, on which date OKB-1 formally embarked upon work related to the program, informal work having commenced prior to this.

The R-36 ICBM (above) was further developed into the 8K69 FOBS, which in turn formed the basis for the Cyclone-2 launch vehicle that would carry the IS Satellite Fighter into orbit. US DoD

While the UR-200 and GR-1 had both fallen by the wayside, Yangel was pushing on with development of a potential global rocket modification of the 8K67 missile of the R-36 ICBM complex and other design houses put forward designs. Following the cancellation of the UR-200, development of a larger launch vehicle had continued in support of the Soviet manned space program for employment in such areas as the launch of Space gliders and rocket gliders. This research led to the initiation of development of the 500 ton class UR-500 launch vehicle designed and developed under V. Chelomey of Affiliation 1 of the Machine Building Central Design Bureau in cooperation with OKB-52. The RD-206 engine of the UR-200 effectively became the prototype of the RD-0208 and RD-0209 engines used in the second stage of the UR-500, which could orbit a payload of up to 20 tons. The development program was directed toward testing two and three stage variants during the period 1965-1968. The maiden launch, which occurred on 16 July 1965, placed a Proton research satellite into LEO (Low Earth Orbit), the UR-500 adopting the name Proton from this first payload carried into orbit.

Flush with the success of the two-stage UR-500 rocket, the design team embarked upon further development of the planned three stage variant, which was aimed at launching manned circumlunar missions under Project LK-1, this emerging as the UR-500K heavy launch vehicle, which, as noted above, would later carry the name Proton, with a take-off mass of 700 metric tons when fitted with the Block D upper stage. Like the UR-200 before it, the UR-500 launch vehicle had been designed with the potential for development as a strategic ballistic missile complex or as a 'Global Rocket' armed with what was termed 'global warheads' that would be carried into orbit and be maneuvered in Earth orbit before being deorbited overs a specific target area. However, the UR-500K, the first launch of which occurred on 10 March 1967, was developed purely as a space payload launch vehicle as the Soviet Union had selected the 8K69 as the basis of its Global Rocket a few years previous.

Two orbital maneuvering warheads designs were developed for the UR-200/UR-500 orbital weapon programs - the AB-200 for use on the IBM UR-200 and the AB-500, which would, it was planned, arm the projected IBM UR-500 missile complex. These warheads were designed in such a manner as to be capable of maneuvering in Earth's atmosphere after de-orbit in order to increase accuracy. To this end, OKB-52 developed the MP-1 warhead vehicle, which was to be capable of using aerodynamic control to maneuver in Earth atmosphere at high hypersonic speeds, this having been accomplished in what was described as a successful test in 1961 when the UR-200 was under development.

Like the IBM UR-200, the IBM UR-500 complex was never seen through to fruition. The IBM UR-200 program had been considered not technically viable whilst the larger IBM UR-500 was dropped in favor of the more, near term, technologically attainable and less costly 8K69 of the R-36 missile complex.

Although the MP-1 manoeuvre vehicle was dropped following the cancellation of the IBM UR-200/500 programs, technology from this program fed directly into the Polyet-1, the world's first space manoeuvring satellite. The Polyet-1, which JSC MIC Mashinostroyenia confirms was effectively the prototype flight system for the IS complex developed by CB-1, was launched on 1 November 1963. This spacecraft demonstrated a rudimentary capability for one spacecraft to converge on the orbital position of another spacecraft, although there was no other spacecraft involved in the actual demonstration. The second flight of a Soviet maneuvering satellite was conducted on 12 April 1964 with the launch of the Polyet-2.

In 1965, CB-1, also referred to as DB-41 or KB-1 (now Open Joint Stock Company 'Corporation Space System Special, Comet) took on the mantle as lead design bureau for the IS program under chief designer A.I. Savin (Vlasko-Ka Vlasov became chief designer on the program in 1979, being succeeded by L.S. Legeza in 1983). Not long after CB-1 assumed the lead role in development of the IS complex, development commenced of a complex of ground control centres – OKO - for operational control of the IS (referred to as the Ka-interceptor), which, as noted above, would be carried into orbit aboard a Cyclone-2 launch vehicle.

Given the technology levels of the time, establishment of a viable space defence system was a major undertaking with many problems to be overcome. Among the areas that required systems development and testing was the following laid down in OJSC Corporation Space System Special, Comet documentation: "detectors and satellite service (later Centre) space control, which laid the foundation for today's space control system (SKKP); rocket and space complex operational start up KA-interceptor; KA-interceptor capable of high precision maneuvering in space to solve the problem of autonomous detection of KA-purpose [KA target] and homing in on it; automated ground control centre systems management".

A cooperation of industrial organisations was set up to implement the program under the umbrella of CB-1 (KB-1, DB-41), working for and in cooperation with the Ministry of Defence of the Soviet Union. Great advances in technology were required in such areas as the target designation systems, which, between the various elements of the command systems, the rocket space complex, the KA-interceptor and the ground-based automated data-processing systems, handled targeting data over distances that would cover many thousands of kilometers in near Earth space.

A US intelligence depiction of a Soviet ASAT launch facility complete with one launch vehicle on the launch pad and others in storage or in the process of being moved by rail. DIA

Cyclone-2 space payload launch vehicle. Yuzhnoye State Design Office

Development of the 8K69 orbital weapon complex had positioned OKB-586 well for development of a lightweight launch vehicle based on the former system. Development of the resultant 11K69 Tsyclone-2 (Cyclone-2) two-stage launch vehicle (allocated the western space launch vehicle designation SL-11) commenced in 1965. The Cyclone-2 was designed to place payloads into circular and elliptical low Earth orbits that would include what was referred to as open orbits of a type that could be beneficial to an IS system. The term open orbit referred to the fact that the satellite would not be restricted to the closed orbit characteristics that basically saw an object such as a satellite effectively return to its start point following each complete orbit of the Earth.

The Cyclone-2 was apparently the first rocket system, certainly within the Soviet Union, designed with what is described in Yuzhnoye State Design Office documentation as a "fully automated launch operations cycle" that required an extensive development period. The Cyclone-2 designs maiden launch was conducted from Baikonur Cosmodrome on 6 August 1969, there being a total of 106 launches of the system during its operational life with the Soviet Union and later the independent Ukraine, the last of which was conducted in June 2006. The Cyclone-2 demonstrated unprecedented and unrivalled reliability with all 106 launches being successful (the value of 106 launches was provided by Yuzhnoye State Design Office, whilst the National Space Agency of Ukraine puts the number of launches at 116, all of which were stated as being successful).

As well as being the launch vehicle for the Counter-Space Defense Satellite Fighter the Cyclone-2 was employed as the launch vehicle of choice for many more traditional payloads, including the RORSAT nuclear powered ocean surveillance satellites.

The limitations of the payload lifting capability of the Cyclone-2 too medium and elliptical Earth orbits led to development of the three-stage 11K68 Cyclone-3 launch vehicle, development of which commenced in 1970 on the basis of the Cyclone-2. However, this latter launch vehicle was not adopted for any element of the Soviet Counter-Space Defence Complex, which did not require the extra stage.

Cyclone-2 Launch Vehicle Specification – data furnished by Yuzhnoye State Design Office with input from the National Space Agency of Ukraine

Number of stages: 2
Length: 39000 mm
Stage diameter: 3000 mm
Main fairing height: 7.0 m
Main fairing diameter: 2.5 m
Payload fairing diameter: 2200 mm
Launch weight (with a 3.2 ton payload): 183 tons
Payload capability: 1.5-5 tons (National Space Agency of Ukraine states 3.2 tonnes maximum)
Propellants
Oxidizer: NTO
Fuel: UDMH (Unsymmetrical Dimethylhydrazine) propellant had a desirable 'quality, spontaneous ignition (hyperbolicity) of the components of the propellant mixture allowing the fueled launch vehicle to stand for extended periods of time. By contrast, first generation ICBM's like the R-7 burned the old mix of kerosene and liquid-oxygen
Vacuum thrust
Stage 1: 303.2 tnf.
Stage 2: 101.5 tnf.
Launch azimuth: 51-99°

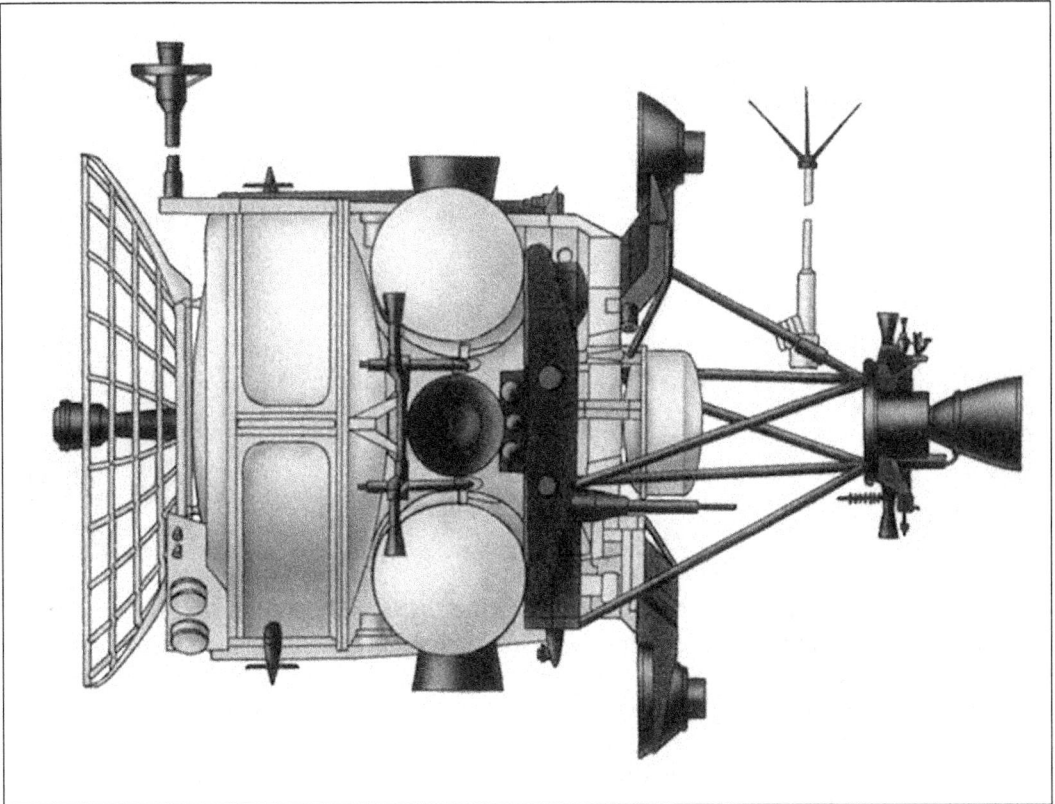

Previous page: The Polyet-1, designed by OKB-52 (JSC MIC Mashinostroyenia), was the words first maneuverable satellite. This spacecraft, which was launched on 1 November 1963, was in effect a prototype for the IS satellite fighter that would emerge several years later. JSC MIC Mashinostroyenia **This page: When referring to the term 'Satellite Fighter' it is easy to allow the mind wander off to exotic looking space planes like this JSC MIC Mashinostroyenia concept of the 1960's (top).** JSC MIC Mashinostroyenia **However, the IS satellite fighter that emerged toward the late 1960's was a far simpler concept (above), albeit advanced for its time.** OJSC 'Corporation' Space System Special, Comet

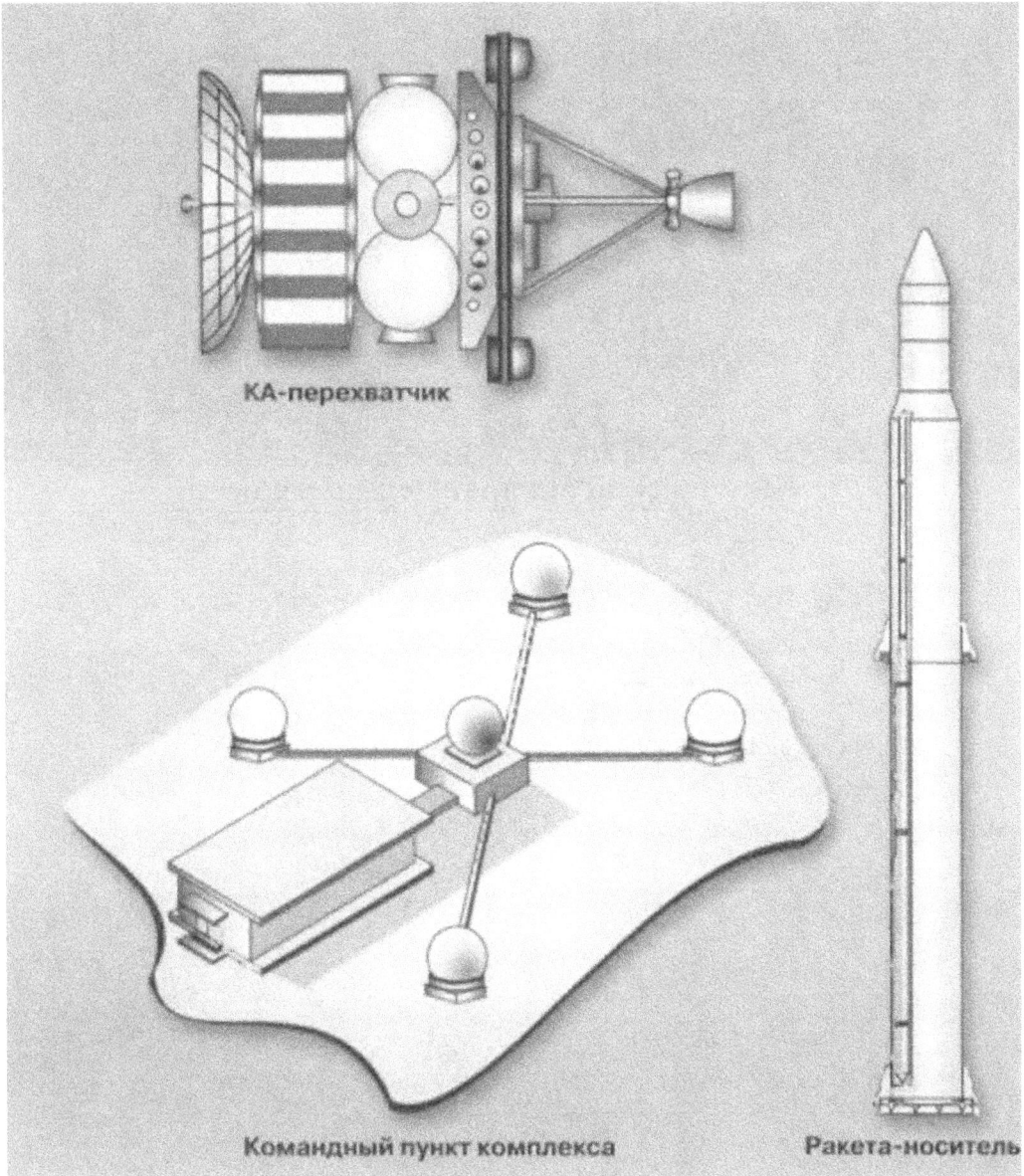

КА-перехватчик

Командный пункт комплекса   Ракета-носитель

Russian language diagram depicting the major elements of the IS Satellite Fighter Complex - Satellite Fighter, Cyclone-2 launch vehicle and ground control stations.
OJSC 'Corporation' Space System Special, Comet

The IS interceptor itself could be described as being cylindrical in appearance with a number of protrusions for various systems - the antenna dish, located at the front of the vehicle, and a propulsion nozzle, located at the rear of the vehicle, being the most prominent. Total mass of the fully equipped and fueled spacecraft was 2450 kg, considerably higher than the mass assumed by western intelligence and scientific organisations – NASA had initially estimated the total mass of the vehicle as 1400 kg.

Космический аппарат-перехватчик

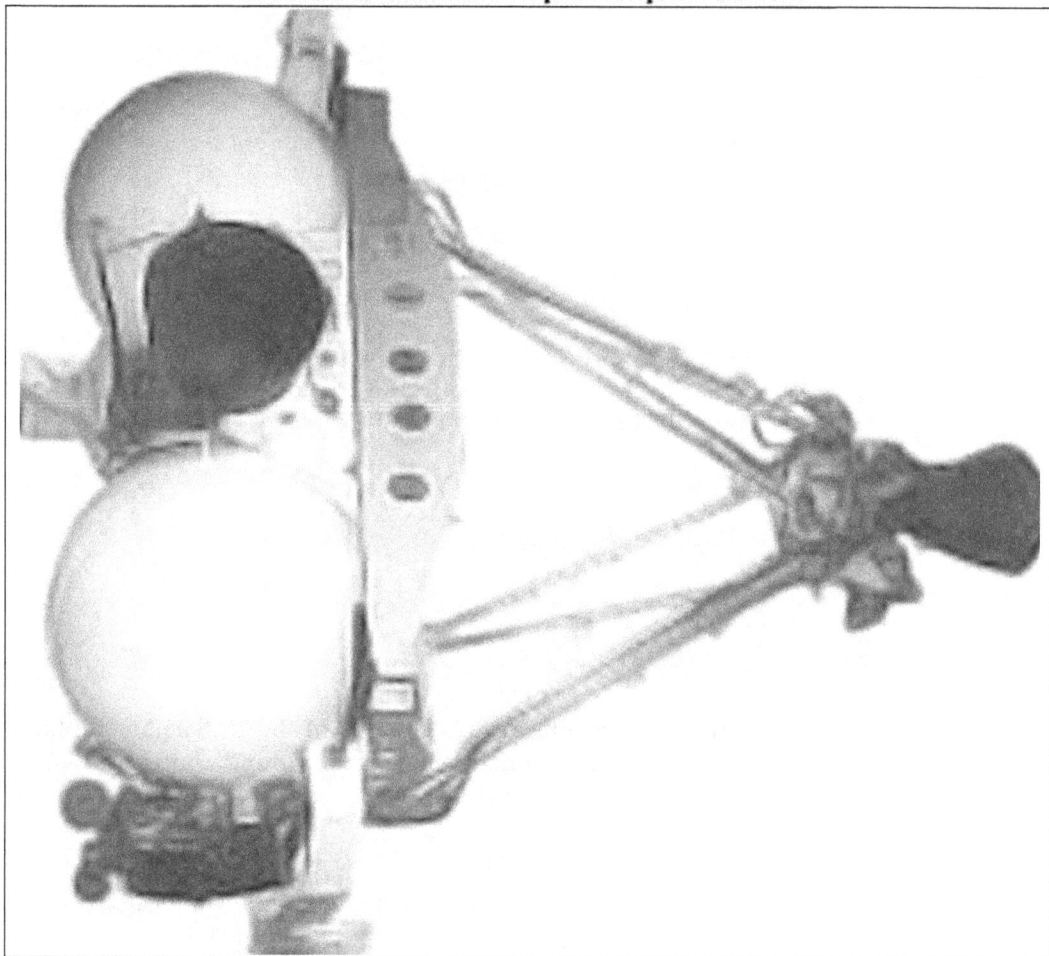

Profile view of an IS Satellite Fighter (top) and profile view of the spacecraft's engine/fuel tank compartments (above). OJSC 'Corporation' Space System Special, Comet/ TMBDB Soyuz

It would be possible to describe the vehicle as being broken down into two main sections, the forward section containing the sophisticated, for the time, guidance, stabilisation and targeting systems, a suite of computational and homing instruments and the hard kill device which contained an explosive fragmentation charge estimated at ~300 kg mass. In 2017, it remains unclear if this particular charge, when employed for the first series of flight tests, was intended to achieve the hard kill or was merely used as a self-destruct mechanism to ensure the destruction of the vehicle at missions end. It was widely assumed that once the explosive charge was detonated that the IS interceptor would itself fragment into a number of groups of shrapnel (NASA (National Aeronautics and Space Administration) documentation puts this at 12 separate groups although it is not clear on what evidence this assessment is based) to achieve a hard kill against an orbiting spacecraft. Effective kill radius was never disclosed by the Soviet Ministry of Defence nor the ministry of defence of the succeeding Russian Federation, but it has been estimated by NASA (presumably through interpretation of such mission parameters that were able to be observed) as being in the region of 1 km. NASA further estimated that the kill radius would be reduced to around 400 km when employed in a head on interception profile, but increased to 2 km when employed in a chase interception profile.

IS Satellite Fighter during testing in the late 1960's/early 1970's. OJSC 'Corporation' Space System Special, Comet

The other main section was concerned with the TMBDB Soyuz developed 5Д18 (5D18) rocket propulsion system. This particular power plant provided spacecraft stabilisation and allowed acceleration when required and maneuvering of the spacecraft as it rendezvoused with the target satellite during the interception process. The engine provided the IS Interceptor vehicle with excellent maneuverability, the engine itself being capable of a number of quick starts and stops to enable the IS to catch up with and close in on a non-cooperative target satellite that was being actively maneuvered to try to avoid interception. Estimated total burn time was around 300 seconds, allowing for a large number of maneuvers to be conducted through the main nozzle and lateral manoeuvre nozzles.

The 5Д18 development program, which effectively kick-started TMBDB Soyuz space programs, incorporated technology advances to solve many of design problems that had dogged early spacecraft. Such measures provided the satellite fighter with an efficient orientation that was significantly increased over systems that had previously been required for spacecraft. The low-thrust of the rocket engine aided spacecraft stabilization and adjustment systems. A unique, for the time, design of spherical fuel tanks featured metal diaphragms and the fuel system provided for the centre-of-gravity of the spacecraft to move as fuel was used up.

Co-Orbital Satellite Fighter specifications – data furnished by TMBDB Soyuz

Initial mass of fueled spacecraft: 2450 kg
Delta-velocity budget: 1200 m/sec
Propulsion system: 5Д18 (5D18) liquid rocket engine capable of multiple ignitions
Service life: 6 years
Launch vehicle: Tsyclon-2 (Cyclone-2)

Whilst the Satellite Fighter (IS interceptor) was launched into Earth orbit atop the Cyclone-2 space payload launch vehicle, the target satellites were launched into Earth orbit on the 11K65M Kosmos-2 (NATO reporting designation SL-8 Mod 1) launch vehicle. This lightweight launch vehicle had emerged from a perceived requirement for a cheaper, less complex, space payload launch vehicle to supplement the OKB-1 (Energia) R-7 ICBM derived heavy launch vehicles that put the first generations of Soviet spacecraft into Earth orbit commencing with the launch of Sputkin-1 in October 1957. OKB-586 had found itself well placed to put forward a design based on the single-stage R-12 IRBM (Intermediate Range Ballistic Missile), which would have a new second-stage added. The resultant lightweight launch vehicle, capable of placing payloads of up to 450 kg into low-Earth orbit, being accepted for production and service.

The first successful flight of a Kosmos launch vehicle took place on 16 March 1962 when the vehicle placed an OK-586 designed DS-2 small size spacecraft into low-Earth orbit. Designated Kosmos-1, the design was operated from the Kapustin Yar and Plesetsk launch facilities, the last of 165 launches (143 of these were successful) taking place on 18 June 1977.

The tandem-stage Kosmos-2 was developed from the R-14 IRBM by incorporating a new design second stage atop the missile. This new lightweight launch vehicle could accommodate clusters of up to 8 small-size spacecraft that would be released in circular low-Earth orbits of altitudes up to 2000 km. As OKB-586 was heavily committed to other projects, in particular the R-36 and R-56 missile programs, further development of the Kosmos-2 was transferred to OKB-10 at Krasnoyarsk in Soviet Russia. A further development of the system resulted in the Kosmos-3M, which was produced under the overall management of Production Association Polyot in Omsk, Russia, from 1970.

The first launch of a Kosmos-2 took place on 18 August 1964, in excess of 700 launches being accomplished in total, with in excess of 1,000 spacecraft delivered into Earth orbit.

The Kosmos family of launch vehicles had the distinction of being the only such systems to be operated from three separate launch sites, Baikonur, Plesetsk and Kapustin Yar.

Kosmos-2 11K65M (SL-8 Mod 1) Launch Vehicle Specification – data furnished by Yuzhnoye State Design Office

Number of stages: 2
Length: 32400 mm
Stage diameter: 2400 mm
Payload fairing diameter: 2400 mm
Launch weight: 109 tons
Propellants
Oxidizer: AK-271
Fuel: UDMH (Unsymmetrical Dimethylhydrazine)
Stage 1 vacuum thrust: 177.5 tnf.
Stage 2 vacuum thrust: 16 tnf.
Payload capability to circular orbits (200-2000 km altitude): up to 1.5 tons

Previous page and above: Kosmos-2 space payload launch vehicle. Yuzhnoye State Design Office

The target spacecraft utilised in the various development and operational test phases of the Soviet ASAT program were developed by the Adjustment and Calibration Spacecraft division of OKB-586 (Yuzhnoye State Design Office). This Division was responsible for the design and development of a number of 'special purpose' spacecraft for the Ministry of Defence of the Soviet Union, including the DS-P1-Yu, DS-P1-I first generation spacecraft and the Tjulpan (Tulip) second generation spacecraft. While the DS-P1-Yu was developed for calibration test operations concerning the Soviet early-warning radar system and the DS-P1-I was employed on functional checkouts of ground stations, the Tjulpan was designed for active employment in development of the operational capabilities of the Soviet ASAT system. The first orbital test flight of a Tjulpan spacecraft, designated Cosmos 394, took place in 9 February 1971 as a target mission for the Cosmos 397 IS interceptor launched just over two weeks later. However, it is known that spacecraft were utilised as ASAT targets as early as 1968. Although confirmation has only been received about the use of the Tjulpan in the ASAT target role, the best available evidence would suggest that either the DS-P1-Yu or the DS-P1-I spacecraft were employed as targets for ASAT tests conducted between 1968 and 1970.

While the Tjulpan second generation target had a total mass of 643 kg, the first generation target being in a similar weight class, western intelligence and scientific organisations had estimated mass at around 4000 kg down to 750 kg, this lower value being the closest to the true mass of 643 kg.

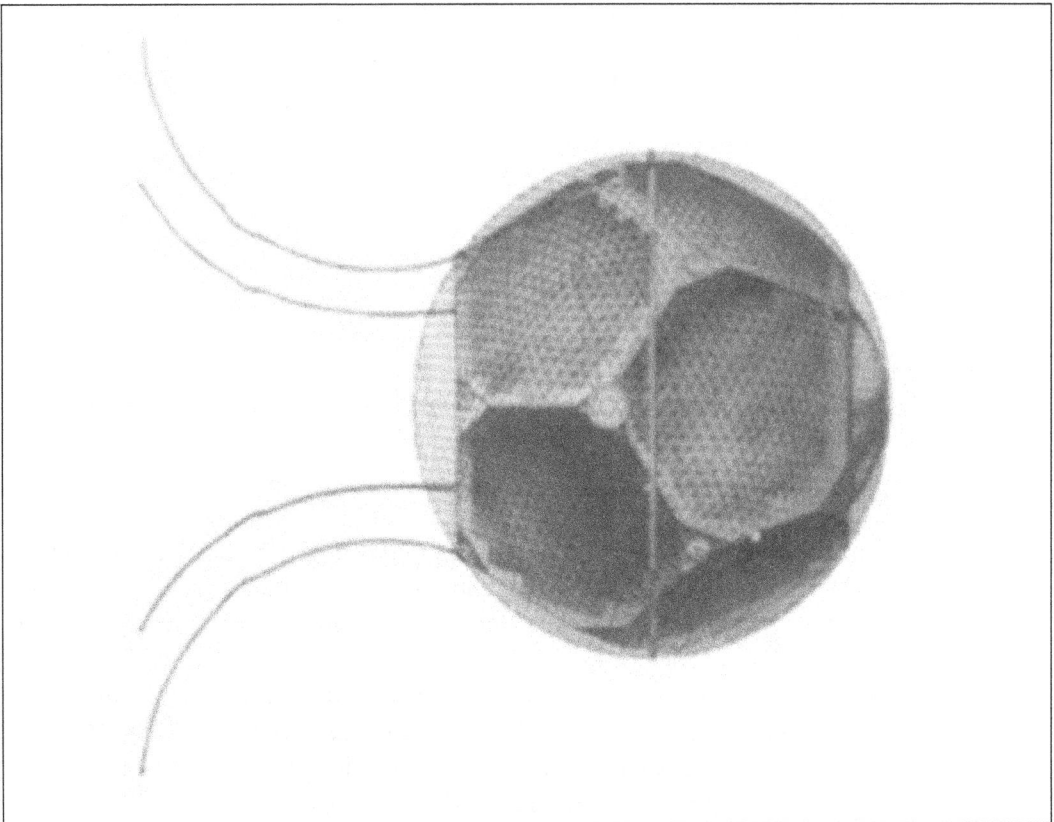

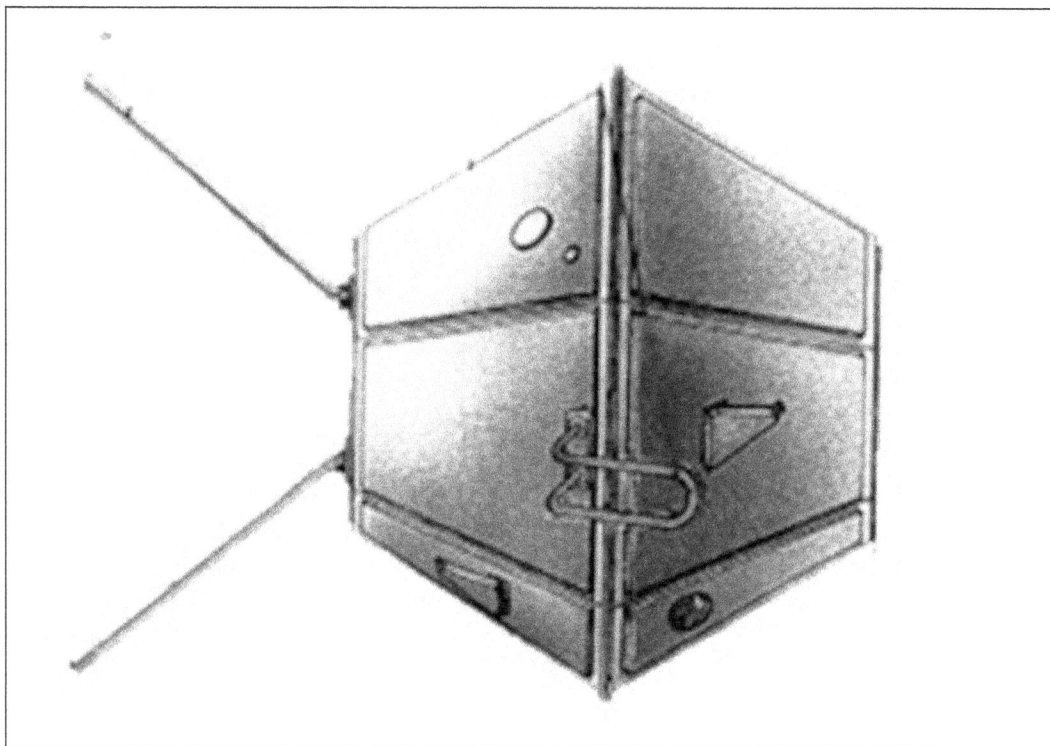

Previous page: DS-P1-YU adjustment and calibration spacecraft, a type thought to have been employed as a target spacecraft in the initial ASAT test phase. Above: The Tjulpan adjustment and calibration spacecraft was employed as a target spacecraft from February 1971. Yuzhnoye State Design Office

Tjulpan specification – data furnished by Yuzhnoye State Design Office

Mass: 643 kg
Active life: 60 days
Launch vehicle: 11K65M
Main experimental results: The Tjulpan spacecraft was a target satellite developed for full-scale tests of killer satellites of the Soviet ant-satellite defence system – the Co-Orbital Satellite Fighter complex
First orbited: 9 February 1971

Command and control of an IS interception mission was commanded from a Ground Command Post that would provide and receive all relevant mission information such as coordinate measurement, command transmissions, reset of telemetric information and trajectory measurements to and from the IS interceptor.

The Soviet command and control network for ICBM and spacecraft in Earth orbit had its early beginnings following the 1955 decision to build an ICBM test facility, now part of what is known as Baikonur Cosmodrome in Soviet Kazakhstan (now an independent republic). This KIK (Ground Control Station) attained operational capability in 1957, the year of the first successful ICBM launch and the first artificial satellite to be placed in Earth orbit. As well as construction of an ICBM test range, in 1960 the 3rd Department of the Main Missile Directorate of the Ministry of Defense of the USSR was formed to facilitate the organisation of space control. In 1964, this organisation would be formed into the Central Spacecraft Directorate of the USSR Defense Ministry (Russian abb, TsUKOS). In 1970, it formed the basis of the Soviet Main Spacecraft Directorate of the USSR Defence Ministry (Russian abb, GUKOS). GUKOS and subordinate organisations were, in 1982, reassigned form the Soviet Strategic Missile Force to come under the direct command of the USSR Defence Minister for a time until the Defence Space Units was formed. In August 1992, following the dissolution of the Soviet Union the previous December, the Russian Federation established its Space Forces, which controlled the Baikonur, Plesetsk and Svobodny cosmodromes and the Main Trial Centre for Testing and Control of Space Means (Russian abb, GITSIU KS), being incorporated from 1994, by which time the Co-orbital satellite fighter had been withdrawn from operational service.

The command and control for the Soviet space defence programs was built around an extensive network of stations, which would include such facilities as the OKO system of the East Command Post. OJSC 'Corporation' Space System Special, Comet

Basic flight profile - An operational launch in times of hostilities could have taken advantage of the systems so called 'multiwind option interception' in that the interceptor could be launched in a northerly or southerly trajectory, the former favouring a single or multiple co-orbital interception profile and the latter, perhaps, benefitting a point interception in less than a complete orbit.

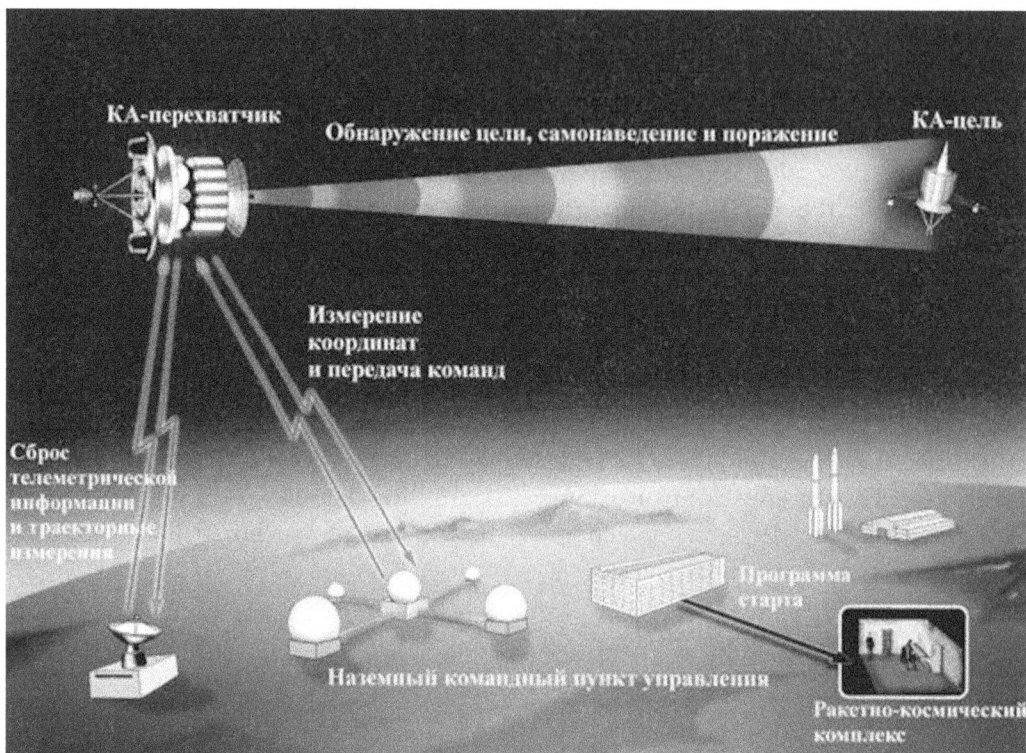

КА-перехватчик Обнаружение цели, самонаведение и поражение КА-цель

Измерение координат и передача команд

Сброс телеметрической информации и траекторные измерения

Программа старта

Наземный командный пункт управления

Ракетно-космический комплекс

Схема действия комплекса "ИС"

This Russian language diagram roughly translates to the KA-Interceptor (ка-перехватчик) conducting 'target detection, self-immolation and defeat [destruction]' обнаружение цели, самонанедение и поражение (obnaruzheniye tseli, samonanedeniye i porazheniye) target (ка–цель – goal) under control of the 'ground command post' (наземный командный пункт управления) for 'coordinate measurement and command transmission' (измерение координат и передача команд) and 'reset telemetric information and trajectory measurements' (сброс телеметрическсй информации и траекторные измерения). Also depicted is the 'rocket space complex (ракетно-космическии комплекс) and the 'launch program' centre (программа старта). OJSC 'Corporation' Space System Special, Comet

The basic flight profile of the satellite fighter would involve the launch atop a Cyclone-2 launch vehicle, commencing the active phase of the operation, which would have followed the pre-launch phase. The spacecraft would then be placed into Earth orbit during the staging phase. This would be followed by an acceleration phase in order that the satellite fighter could reach the open orbit required to attempt an interception of the target spacecraft. During this phase any required trajectory corrections would have been implemented as it was conducted concurrently with the so called 'finding-search-field' phase as the satellite fighter attempted to acquire the target to facilitate a rendezvous. Once the rendezvous had been attained the interception was conducted at the closest practical approach. A successful interception would result in a target being destroyed or damaged to the point it was *hors de-combat* (no longer mission effective).

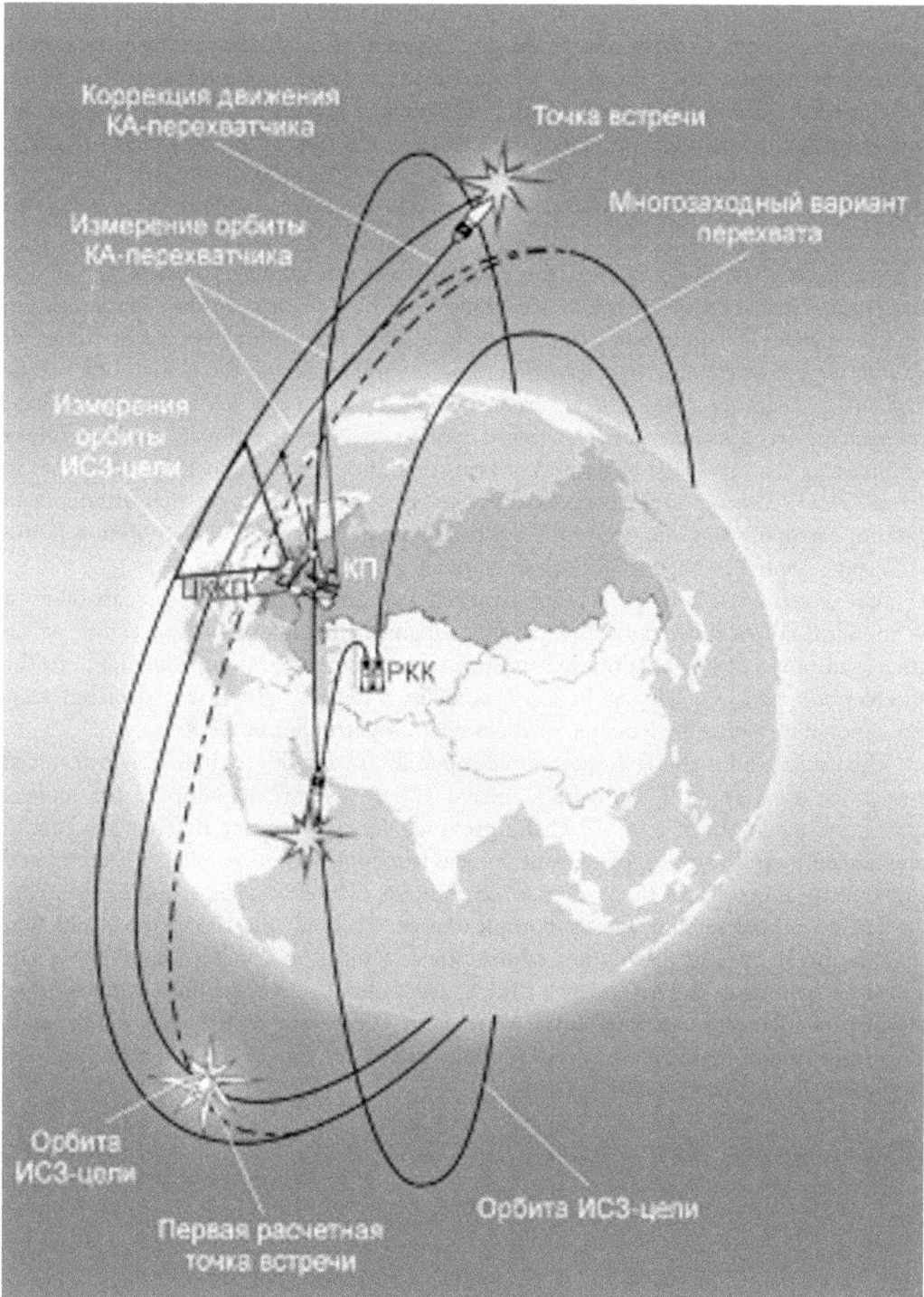

Russian language diagram that shows the basic launch, point (fraction of an orbit) and full orbital interception profiles for the Co-Orbital Satellite Fighter. OJSC 'Corporation' Space System Special, Comet

An interception would be considered successful under a number of scenarios, not only the destruction of the target satellite. Even if the satellite was not destroyed, perhaps by maneuvering out of its orbit in an attempt to avoid interception, it would be rendered mission ineffective as such a maneuver would have to be extreme and would probably only delay interception if the satellite fighter mission was continued. This, of course, could, in such circumstances, have been deemed unnecessary as the target satellite would be *hors de combat* in terms of mission availability.

The initial test phase to clear the co-orbital satellite fighter complex for limited operational capability consisted, according to OJSC 'Corporation' Space System Special, Comet documentation, of 7 full-scale flight tests. However, eight flights of IS interceptor spacecraft were recorded between and including the launch of Cosmos 185 on 27 October 1967 and Cosmos 462 on 3 December 1971. This test phase proved the basic interception capability, paving the way for the IS satellite fighter complex to be introduced to limited or trials operational service in 1972 (information from OJSC 'Corporation' Space System Special, Comet indicates that although an interim capability was achieved in 1972, the Counter-Space Defence System was not officially adopted by the USSR Supreme Soviet until 1973).

A renewed test effort from February 1976 was aimed at proving the capability of a modernised satellite fighter with the capability to intercept spacecraft at an increased range of orbital altitudes and orbital inclinations and improvements to the interception process. This enhanced capability satellite fighter was declared fully operational in service by a decree of the Soviet Government in 1978.

The same year as the full operational capability IS satellite fighter was introduced to service the Ground Command Centre – OKO system was adopted for service, state tests having been completed the previous year, as a space defence system for the detection and tracking of ballistic missile launches and, in parallel, for command and control functions of the rocket-space complex of the IS satellite fighter complex.

The co-orbital satellite fighter complex served through the rest of the Cold War and the break-up and dissolution of the Soviet Union on 25 December 1991. The system remained in the inventory of the space-defence forces of the newly emerged Russian Federation, serving until retired in 1993 when the last units were decommissioned, ending the operational career of the world's only dedicated operationally deployed ASAT system.

3

CO-ORBITAL SATELLITE FIGHTER DEVELOPMENT AND OPERATIONAL TESTING – 1963-1982

There are 39 spacecraft flights that can be directly attributed to the Soviet Counter-Space Defence Satellite fighter (IS interceptor) Complex development and operational test programs. The first spacecraft that can positively be attributed to the development of an ASAT (Anti-Satellite) system was the OKB-52 (JSC MIC Mashinostroyenia) Polyet-1 manoeuvrable spacecraft, which was launched in 1963. This spacecraft, which effectively acted as a prototype for what would become the IS interceptor, was developed from experience gained in development of the MP-1 maneuvering in the atmosphere high hypersonic vehicle that was being developed under the Soviet Union's orbital weapons programs. Whilst the Polyet-1 and Polyet-2 were launched aboard R-7 derived launch vehicles all other spacecraft launched under the ASAT development program were launched on Cyclone-2 (IS interceptors) or Kosmos-2 (calibration and target spacecraft) launch vehicles.

A maneuverable satellite launch failure on 25 January 1969 was recorded in intelligence document NIE 11-1-69, *'The Soviet Space Program'*, although this is not replicated in other records either in Russia or at NASA (National Aeronautics and Space Administration) from where the intelligence data apparently originated. The only Soviet space launch recorded on the date in question was Cosmos 266, a photo reconnaissance satellite launched on a Soyuz launch vehicle from Plesetsk cosmodrome. Therefore, the alleged 25 January 1969 launch failure is not entered in the chronological listing below. Other calibration spacecraft were employed in the development of the Soviet Union's ICBM detection/near-Earth space monitoring and control complexes, but these, being incidental, are not recorded here.

At the same time as development of the IS system other maneuverable satellites were being developed in the Soviet Union for other roles such as reconnaissance, in particular the nuclear powered RORSAT (Radar Ocean Reconnaissance Satellite), which was designed to search vast ocean areas for NATO (North Atlantic Treaty Organisation) shipping from a low-Earth orbit. The first of the RORSAT's, Cosmos 209, was launched atop a Cyclone-2 launch vehicle on 22 March 1968.

By early 1968, western intelligence agencies, having had time to digest available evidence, were still divided as to the exact purpose of the Soviet maneuverable satellite programs, although it is clear from available intelligence documents that the finger of suspicion was motioning in the direction of an ASAT system. Intelligence document NIE 11-1-67 noted that the tempo of work on Soviet maneuverable satellite programs had increased. Two satellites were noted to be employed on maneuverable programs, one estimated at 4082 kg (9,000 lb.) mass and the other at 3175 kg (7,000 lb.). Although estimates of spacecraft mass were off by a considerable margin, almost double for the first and in the region of four times as large for the second type of spacecraft, it was becoming clear that the Soviets may be testing potential ASAT technologies, although a number of other potential roles were also identified. The evidence for this emerged from a series of manoeuvring satellite testing that appeared to be aimed at developing the ability to direct one spacecraft to pass within close proximity of passive and non-cooperating target satellites.

The above inference came from a series of missions that began with the launching of Cosmos 248 on 19 October 1968. This spacecraft, it was noted from orbital data, appeared to be acting as a rendezvous target for Cosmos 249 and 252, launched on 20 October and 1 November 1968 respectively. On the second orbit of Cosmos 249, the spacecraft maneuvered to pass within 70 miles of the Cosmos 248 target. During its third orbit Cosmos 249 was noted to be "tumbling or spinning and was accompanied by several fragments" (NIE 11-1-67), while Cosmos 248 remained in its orbit, enabling the spacecraft to be employed as a target for Cosmos 252, which, on its second orbit, was maneuvered to pass within 1 mile of Cosmos 248. It was noted in intelligence document NIE 11-1-67, "Unlike the previous operation, in this instance both Cosmos 248 and 252 were accompanied by fragments after the operation; our calculations indicate that this fragmentation occurred at about the time of the fly-by. Several orbits later, both vehicles were observed tumbling or spinning". This led to an inference among in the intelligence community that Cosmos 252 had actively sacrificed itself to destroy Cosmos 248.

Although ASAT capabilities were one school of thought, statements circulated indicated that opinion was divided on potential roles for the maneuverable spacecraft being tested by the Soviet Union. One such statement within the text of intelligence document NIE 11-1-67 read "While we cannot determine the missions of the various satellite's involved, the maneuverable satellite program could be intended to fulfill a number of roles, both military and non-military. We are not yet able to determine the most likely roles that will evolve. Some flights could be directed solely towards the development of a multi-purpose orbital propulsion capability, the techniques of which could support a variety of intercept or rendezvous mission's. A close fly-by at high relative velocity would be a requirement for one form of an anti-satellite system. Alternatively, a close fly-by, but at lower relative velocity could indicate an intent to rendezvous or fulfill an inspection mission [It was noted that during the above noted fly-by missions, high-speed closure rates of some 365.76 m/s (1,200 fps) were recorded]. We believe, however, that the Cosmos 248, 249 and 252 operation is more applicable to an anti-satellite role that any other mission objective". This assessment was in the ball park in regards to the Cosmos 248, 249 and 252 missions, however, as noted above, the Soviet Union was pursuing duel maneuverable satellite programs to facilitate the acquisition of an ASAT complex and its ocean surveillance satellite requirements.

Polyet, IS (Satellite Fighter)/target ASAT flight chronology					
Vehicle	Type	Launch date	Destruction	Decay	Deorbit
Polyet-1	Demo	01.11.1963			
Polyet-2	Demo	12.4.1964			
Cosmos 185	IS	27.10.1967		14.01.1967	
Cosmos 217	Target	24.4.1968		26.04.1968	
Cosmos 248	Target	19.10.1968	01.11.1968	26.02.1980	
Cosmos 249	IS	20.10.1968			
Cosmos 252	IS	01.11.1968			
Cosmos 373	Target	20.10.1970		08.03.1980	
Cosmos 374	IS	23.10.1970			
Cosmos 375	IS	30.10.1970			
Cosmos 394	Target	09.02.1971			
Cosmos 397	IS	25.02.1971			
Cosmos 400	Target	19.03.1968			
Cosmos 404	IS	04.04.1971			
Cosmos 459	Target	29.11.1971		27.12.1971	
Cosmos 462	IS	03.12.1971		04.04.1975	
Cosmos 521	Target	29.09.1972			
Cosmos 803	Target	12.02.1976			
Cosmos 804	IS	16.02.1976		16.02.1976	
Cosmos 814	IS	13.04.1976			13.04.1976
Cosmos 839	Target	09.07.1976			
Cosmos 843	IS	21.07.1976			21.07.1976
Cosmos 880	Target	09.12.1976		31.08.1991	
Cosmos 886	IS	27.12.1976			27.12.1976
Cosmos 909	Target	19.05.1977			
Cosmos 910	IS	23.05.1977			23.05.1977
Cosmos 918	IS	17.06.1977			18.07.1977
Cosmos 959	Target	21.10.1977		30.11.1977	
Cosmos 961	IS	26.10.1977			26.10.1977
Cosmos 967	Target	13.12.1977			
Cosmos 970	IS	21.12.1977			
Cosmos 1009	IS	19.05.1978			19.05.1977
Cosmos 1171	Target	03.04.1980			
Cosmos 1174	IS	18.04.1980			
Cosmos 1241	Target	21.01.1981			
Cosmos 1243	IS	02.02.1981			02.02.1981
Cosmos 1258	IS	14.03.1981			15.03.1981
Cosmos 1375	Target	06.06.1982			
Cosmos 1379	IS	18.06.1982			18.06.1982

Polyet-1: Polyet-1 (confirmed as the flight prototype of the IS complex by designer JSC MIC Mashinostroyenia and OJSC 'Corporation' Space System Special, Comet), the world's first space maneuvering satellite, drew on experience gained in the development of the MP-1, maneuvering in the atmosphere, high hypersonic vehicle that was developed under the Soviet Union's orbital weapons programs. The Polyet-1, launched on 1 November 1963, demonstrated a rudimentary capability for one spacecraft to converge on the orbital position of another spacecraft, although there was no other spacecraft involved in the demonstration. During the mission the spacecraft was able to alter its inclination, apogee and perigee.

The mission had an epoch start of 08:52:00 UTC on 1 November 1963 (conflicting NASA documentation suggests an epoch start of 03.52 UTC on 1 November 1964). The mission orbital parameters included a periapsis of 339 km, apoapsis of 592 km, period 93.3 minutes, inclination 58.8° and an eccentricity of 0.01847.

Polyet-2: OJSC 'Corporation' Space System Special, Comet documentation confirms the launch of Polyet-2 on 12 April 1964. This was the second test flight of the manoeuvrable satellite development program and, like the Polyet-1 spacecraft, was equipped with control systems that allowed it to alter its inclination, apogee and perigee.

The mission had an epoch start of 09:21:00 UTC on 12 April 1964 (conflicting NASA documentation suggests epoch start of 04.21 UTC on 12 April 1964). The mission orbital parameters included a periapsis of 242 km, apoapsis of 485 km, period 92.4 minutes, inclination 58.1° and an eccentricity of 0.01801.

Cosmos 185: Cosmos 185, which was launched from the Baikonur cosmodrome at Tyuratam in Soviet Kazakhstan on board a Cyclone-2 two-stage launch vehicle on 27 October 1976, was the maiden mission of the IS interceptor (Satellite Fighter) complex as well as the first launch of the Cyclone-2 launch vehicle. Once placed into a low orbit by the Cyclone-2, the IS interceptor, along with the upper-stage section of the Cyclone-2, was maneuvered into a higher orbit after which the upper-stage was discarded.

This mission, which was a basic flight-test of the IS interceptor vehicle, intended to check-out the spacecraft flight systems and communications with the ground-control stations whilst operating in a space environment, had an epoch start of 02:24:00 UTC on 27 October 1967. The mission orbital parameters included a periapsis of 518 km, apoapsis of 873 km, period 98.7 minutes, inclination 64.1° and an eccentricity of 0.02508. Following the tests the spacecraft remained in orbit until the orbit decayed on 14 January 1969.

Cosmos 217: Cosmos 217 was an adjustment and calibration spacecraft employed as a target for ASAT testing. The spacecraft was launched on board a Cyclone-2 two-stage launch vehicle from the Baikonur cosmodrome at Tyuratam in Soviet Kazakhstan on 24 April 1968. The spacecraft failed to separate from the upper stage of the launch vehicle resulting in failure of the mission.

The mission had an epoch start time/date of 16:04:00 UTC on 24 April 1968. The mission orbital parameters included a periapsis of 144 km, apoapsis of 262 km, period 93.4 minutes, inclination 62.2° and an eccentricity of 0.00896. The spacecraft orbit decayed on 26 April 1968.

Cosmos 248: Cosmos 248 was an adjustment and calibration target spacecraft for ASAT testing. The spacecraft was launched into low-Earth orbit on board a Kosmos-2 two-stage launch vehicle from the Baikonur cosmodrome at Tyuratam in Soviet Kazakhstan (conflicting NASA documentation states the launch vehicle was a Cyclone-2) on 19 October 1968. The spacecraft acted as a target for the Cosmos 249 IS interceptor and later the Cosmos 252 IS interceptor, the latter vehicle destroying Cosmos 248 on 1 November 1968.

The mission had an epoch start time/date of 04:19:00 UTC on 19 October 1968. The mission orbital parameters included a periapsis of 475 km, apoapsis of 543 km, period 94.8 minutes, inclination 62.3° and an eccentricity of 0.00493. The orbit of the remains of the spacecraft decayed on 26 February 1980.

Cosmos 249: Cosmos 249 was an IS interceptor spacecraft that was launched on board a Cyclone-2 two-stage launch vehicle from the Baikonur cosmodrome at Tyuratam in Soviet Kazakhstan on 20 October 1968. The spacecraft was placed into a low-Earth orbit, following which it left behind some staging debris in a similar orbit to that of the Cosmos 248 target spacecraft that had been launched the day before, as it transferred to a higher, more eccentric orbit with a perigee similar in altitude to the circular orbit of Cosmos 248, but an apogee around 4 times higher in altitude. Having a perigee similar to the circular orbit altitude of Cosmos 248 allowed the Cosmos 249 interceptor to pass fairly close to Cosmos 248 in a demonstration of intercept profiles. It is considered that a number of such close proximity interceptions were conducted on 20 October 1968, this certainly being the inference of information furnished in NASA documentation, available from monitoring the spacecraft passes. Following the final interception pass of the test phase the Cosmos 249 interceptor moved away from the Cosmos 248 target before the former was exploded, destroying the spacecraft.

The mission had an epoch start date of 20 October 1968. The mission orbital parameters included a periapsis of 517 km, apoapsis of 2092 km, period 111.8 minutes, inclination 62.4° and an eccentricity of 0.10244. The orbit of the remains of the spacecraft decayed on 26 February 1980.

Cosmos 252: The Cosmos 249 mission had allowed the development test team to go through a number of missions manoeuvres required for the IS interceptor mission, but had left the target, Cosmos 248, intact, allowing it be utilised in the Cosmos 252 IS interceptor test flight that was launched on board a Cyclone-2 launch vehicle from Baikonur cosmodrome at Tyuratam in Soviet Kazakhstan on 1 November 1968. Cosmos 252 was intended to be a similar test of the IS vehicle capabilities as had been conducted by Cosmos 249.

As had been the case with the Cosmos 249 interceptor mission, Cosmos 252 was placed into low-Earth orbit where it abandoned staging debris before it was maneuvered into a considerably higher eccentric orbit. As had been the case with Cosmos 249, the perigee of this eccentric orbit was similar in altitude to that of the near circular orbit of the Cosmos 248 adjustment and calibration target spacecraft. This facilitated the passing of the two spacecraft at a fairly close proximity. Western intelligence assessments conclude that such maneuvers were conducted at least once, although data furnished by OJSC 'Corporation' Space System Special, Comet indicates that the interception took place on the second orbit. The warhead carried (OJSC 'Corporation' Space System Special, Comet confirms that the IS was equipped with a fragmentation warhead on the 1 November 1968 flight) on-board Cosmos 252 was exploded, destroying the Cosmos 248 target. OJSC 'Corporation' Space System Special, Comet and JSC MIC Mashinostroyenia confirmed Cosmos 252 as the first successful interception of a satellite target.

The mission had an epoch start time/date of 00:28:00 UTC on 1 November 1968. The mission orbital parameters included a periapsis of 551 km, apoapsis of 2102 km, period 112.2 minutes, inclination 62.3° and an eccentricity of 0.10059. There is no reliable information as to a decay date for the remains of the spacecraft.

In the 1980's this intelligence assessment depiction of an IS intercepting a target satellite was released by the US Defence Intelligence Agency. It depicts a fragmentation warhead being activated to destroy the target. DIA

The following is an English language translation of a OJSC 'Corporation' Space System Special, Comet statement on the Cosmos 252 mission: "The system 'IP' [Istrebitel Sputnik (IS interceptor)] after receiving the targeting ensured high efficiency solutions spacecraft to intercept the task of preparing for the start of the carrier rocket with the spacecraft interceptor was performed for 1 hour, and this, despite the fact that it was necessary to perform an enormous amount of work. The solution of these problems with a given urgency was only possible due to maximum automation of all basic processes of preparation for the launch of space launch vehicle. Reached at the time efficiency had no analogues in the world. Interception of a target spacecraft carried out on the second filter coil [inferred as orbit] SC interceptor. In this means the command post system automatically performed measuring the parameters of its orbit and in view of clarification of the orbit of spacecraft-target according to TSKKP provided by the calculation and the transmission on board spacecraft interceptor."

Cosmos 373: Cosmos 373 was an adjustment and calibration spacecraft target for ASAT testing. The spacecraft was launched into low-Earth orbit on board a Cyclone-2 two-stage launch vehicle from the Baikonur cosmodrome at Tyuratam in Soviet Kazakhstan on 20 October 1970. The spacecraft acted as a target for the Cosmos 374 IS interceptor spacecraft and later the Cosmos 375 IS interceptor spacecraft that were launched on 23 October 1970 and 30 October 1970 respectively. Cosmos 373 was initially placed into a higher than normal eccentric orbit before it separated from the launch vehicle staging section. The spacecraft then maneuvered to a more circular lower altitude orbit that was more in line with the orbits of previous ASAT targets.

The mission had an epoch start time/date of 05:44:00 UTC on 20 October 1970. The mission orbital parameters included a periapsis of 480 km, apoapsis of 553 km, period 94.8 minutes, inclination 62.9° and an eccentricity of 0.00456. The spacecraft remained in Earth orbit until orbital decay occurred on 8 March 1980.

Cosmos 374: Cosmos 374 was an IS interceptor spacecraft that was launched into low-Earth orbit on board a Cyclone-2 two-stage launch vehicle from the Baikonur cosmodrome at Tyuratam in Soviet Kazakhstan on 23 October 1970. The spacecraft was initially placed into an eccentric orbit at the initial orbital altitude of the Cosmos 373 before being maneuvered into a higher altitude, more eccentric orbit. Much of the data available for this space flight comes from NASA, which received the data from undisclosed British sources. This data showed that the Cosmos 374 IS interceptor passed very close to the Cosmos 373 adjustment and calibration target spacecraft when Cosmos 374's perigee was very close to that of the almost circular orbit of Cosmos 373. It is unclear how many such passes were conducted, but certainly at least one occurred. When Cosmos 374 was manoeuvred away from Cosmos 373 later on 23 October the interceptors fragmentation warhead was exploded, resulting in a number of debris groups.

The mission had an epoch start time/date of 04:19:00 UTC on 23 October 1970. The mission orbital parameters included a periapsis of 536 km, apoapsis of 2153 km, period 112.3 minutes, inclination 63° and an eccentricity of 0.10463. There is no reliable information as to a decay date for the remains of the spacecraft.

Cosmos 375: On 30 October 1970, one week after the launch of Cosmos 374, another IS interceptor, Cosmos 375, was launched on board a Cyclone-2 two-stage launch vehicle from the Baikonur cosmodrome at Tyuratam in Soviet Kazakhstan. As had been the case with Cosmos 374, Cosmos 375 was maneuvered from its initial insertion orbit to a higher altitude eccentric orbit that allowed the spacecraft to conduct close passes of the Cosmos 373 target spacecraft when Cosmos 375's perigee was close to that of the almost circular orbit altitude of the Cosmos 373 target. The Cosmos 375 fragmentation warhead was exploded, the spacecraft fragmenting into several debris groups, demonstrating the hard kill procedure for the interceptor.

The Cosmos 375/373 mission was the culmination of the test phase for the first generation IS interceptor/adjustment and calibration spacecraft ASAT tests designed to prove the operational utility of the concept. The official Soviet statement carried by the Soviet Tass news agency stated "… the scientific research envisaged by the program has been completed."

The mission had an epoch start time/date of 02:09:00 UTC on 30 October 1970. The mission orbital parameters included a periapsis of 538 km, apoapsis of 2164 km, period 112.4 minutes, inclination 63°and an eccentricity of 0.10512. There is no reliable information as to orbital decay date for the remains of the spacecraft.

Cosmos 394: Cosmos 394 was the lead mission in the second series of major development testing of the Co-orbital Counter-Space Defence complex as the Soviet Union pushed towards attaining a limited operational ASAT capability. Cosmos 394, which was the first operational use of the OKB-586 designed Tjulpan adjustment and calibration ASAT target spacecraft, was launched on board an 1165M Kosmos-2 two-stage launch vehicle from the Plesetsk cosmodrome in Soviet Russia on 9 February 1971.

The mission had an epoch start time/date of 18:49:00 UTC on 9 February 1971. The mission orbital parameters included a periapsis of 574 km, apoapsis of 619 km, period 96.5 minutes, inclination 65.9° and an eccentricity of 0.00322. There is no reliable information as to orbital decay date.

Cosmos 397: Cosmos 397 was an IS interceptor spacecraft launched on board a Cyclone-2 two-stage launch vehicle from the Baikonur cosmodrome at Tyuratam in Soviet Kazakhstan on 25 February 1971. The spacecraft was placed into an initial orbit before moving to a higher altitude orbit, leaving behind the rocket staging section in the lower orbit. The mission of this spacecraft was to conduct a close pass interception of the Cosmos 395 target launched on 9 February. Following completion of the interception procedure Cosmos 397 moved away from Cosmos 394 before the fragmentation warhead was remote exploded, the spacecraft disintegrating into several debris groups.

The mission had an epoch start time/date of 11:16:00 UTC on 25 February 1971. The mission orbital parameters included a periapsis of 593 km, apoapsis of 2317 km, period 114.7 minutes, inclination 65.8° and an eccentricity of 0.10998. There is no reliable information as to orbital decay date.

Cosmos 400: Cosmos 400 was a Tjulpan adjustment and calibration ASAT target spacecraft launched on board an 1165M Kosmos-2 two-stage launch vehicle from the Plesetsk cosmodrome in Soviet Russia on 19 March 1971. This mission differed from previous ASAT test related missions in that the target was boosted to an orbit in the order of twice the altitude of previous ASAT target tests.

The mission had an epoch start time/date of 21:45:00 UTC on 19 March 1971. The mission orbital parameters included a periapsis of 995 km, apoapsis of 1016 km, period 105 minutes, inclination 65.8° and an eccentricity of 0.00142. There is no reliable information as to orbital decay date.

Cosmos 404: Cosmos 404 was an IS interceptor spacecraft that was launched on board a Cyclone-2 two-stage launch vehicle from the Baikonur cosmodrome at Tyuratam in Soviet Kazakhstan on 4 April 1971. The spacecraft mission was to intercept the Cosmos 400 ASAT target that had been launched on 19 March 1971.

Following insertion into a low-Earth orbit, Cosmos 404 manoeuvred to a higher altitude orbit on the same plane as that of Cosmos 400. Having similar relative velocities, Cosmos 404 and Cosmos 400 were able to remain in close proximity to each other during their proximity passes for a longer period of time than had thus far been attained in any previous test.

The mission had an epoch start time/date of 14:24:00 UTC on 4 April 1971. The mission orbital parameters included a periapsis of 811 km, apoapsis of 1009 km, period 103 minutes, inclination 65.9° and an eccentricity of 0.01357. When the co-orbital interception test phase was completed the Cosmos 404 spacecraft was manoeuvred into a new orbit that would very rapidly decay later on 4 April, the largest pieces of debris impacting Oceanic areas on Earth.

Cosmos 459: Cosmos 459 was a Tjulpan adjustment and calibration ASAT target spacecraft that was launched on board an 1165M Kosmos-2 two-stage launch vehicle from the Plesetsk cosmodrome in Russia on 29 November 1971. The spacecraft mission was to act as a target for the Cosmos 462 IS interceptor spacecraft that would be launched on 3 December 1971. This particular mission was aimed at demonstrating a considerably lower orbital altitude satellite interception than had previously been conducted.

The mission had an epoch start time/date of 17:30:00 UTC on 29 November 1971. The mission orbital parameters included a periapsis of 226 km, apoapsis of 277 km, period 89.4 minutes, inclination 65.8° and an eccentricity of 0.00384. The spacecraft orbit decayed on 27 December 1971.

Cosmos 462: Cosmos 462 was an IS interceptor spacecraft that was launched on board a Cyclone-2 two-stage launch vehicle from the Baikonur cosmodrome at

Tyuratam in Soviet Kazakhstan on 3 December 1971. The spacecraft mission was to intercept the Cosmos 459 ASAT target that had been launched on 29 November 1971.

The spacecraft conducted a series of interception profile manoeuvres before the fragmentation warhead was remote exploded, producing a debris/projectile grouping that destroyed the target, Cosmos 459. The best available observational data suggests that this particular explosion produced only 27 major fragments, a lower number than had been produced in any previous test.

The mission had an epoch start time/date of 12:12:00 UTC on 3 December 1971. The mission orbital parameters included a periapsis of 237 km, apoapsis of 1840 km, period 105.7 minutes, inclination 65.8° and an eccentricity of 0.108. The orbit of the remains of the spacecraft decayed on 4 April 1975.

The successful outcome of the Cosmos 459/Cosmos 462 test concluded the initial Soviet State Testing of the Counter-Space Defence Satellite Fighter complex, clearing the way for adoption by the Soviet Armed Forces as the World's only operationally deployed strategic defence ASAT complex.

Cosmos 521: Cosmos 521 was a Tjulpan adjustment and calibration ASAT target spacecraft that was launched on board an 1165M Kosmos-2 two-stage launch vehicle from the Plesetsk cosmodrome in Russia on 29 September 1972. This particular spacecraft mission was to act as a target for a planned flight test of a modified IS interceptor spacecraft design. However, the launch of the interceptor spacecraft was cancelled, apparently due to a perceived problem with the targeting telemetry system.

The mission had an epoch start time/date of 20:19:00 UTC on 29 September 1972. The mission orbital parameters included a periapsis of 973 km, apoapsis of 1030 km, period 105 minutes, inclination 65.8° and an eccentricity of 0.00386.

Following the Cosmos 521 flight the Soviet Union halted ASAT testing as a unilateral gesture to NATO, having signed, along with the United States, the SALT (Strategic Arms Limitation Treaty) in May 1972. There followed a hiatus of more than three and a half years before the next series of Soviet ASAT testing commenced in early 1976. This series was aimed at fielding the improved second generation Co-Orbital Satellite Fighter that was intended to be introduced to full operational service prior to the ABM Soviet/American (Anti-Ballistic Missile) treaty coming into force, then projected for some time in 1978.

Cosmos 803: Cosmos 803 was a Tjulpan adjustment and calibration ASAT target spacecraft that was launched on board an 1165M Kosmos-2 two-stage launch vehicle from the Baikonur cosmodrome in Soviet Kazakhstan on 12 February 1976. This launch signaled the commencement of a new flight test effort to demonstrate improvements incorporated into the second generation Counter-Space Defence Co-Orbital Satellite Fighter (IS interceptor).

The mission had an epoch start time/date of 12:57:00 UTC on 12 February 1976 (conflicting NASA documentation also states an epoch start time/date of 19.00 UTC on 11 February 1976). The mission orbital parameters included a periapsis of 554 km, apoapsis of 624 km, period 96.4 minutes, inclination 66° and an eccentricity of 0.00502.

Cosmos 804: Cosmos 804 was a second generation IS interceptor spacecraft that was launched on board a Cyclone-2 two-stage launch vehicle from the Baikonur cosmodrome at Tyuratam in Soviet Kazakhstan on 16 February 1976. The spacecraft was maneuvered to conduct a co-orbital interception of the Cosmos 803 target spacecraft, after which it moved away from the target before the fragmentation warhead was exploded.

The mission had an epoch start time/date of 08:24:00 UTC on 16 February 1976. The mission orbital parameters included a periapsis of 149 km, apoapsis of 698 km, period 92.8 minutes, inclination 65.1° and an eccentricity of 0.04033. Following detonation of the fragmentation warhead the orbit of the remains of the spacecraft decayed later on 16 February.

Cosmos 814: Cosmos 814 was a second generation IS interceptor that was launched on board a Cyclone-2 two-stage launch vehicle from the Baikonur cosmodrome at Tyuratam in Soviet Kazakhstan on 13 April 1976. The spacecraft mission was to intercept the Cosmos 803 Tjulpan adjustment and calibration ASAT target that had been launched back in February 1976. This was the first mission that saw the interceptor de-orbited at the end of the mission rather than being exploded.

The mission had an epoch start time/date of 17:16:00 UTC on 13 April 1976. The mission orbital parameters included a periapsis of 150 km, apoapsis of 474 km, period 90.6 minutes, inclination 65.1° and an eccentricity of 0.0242. Following the interception manoeuvres the spacecraft was de-orbited later on 13 April 1976.

Cosmos 839: Cosmos 839 was a Tjulpan adjustment and calibration ASAT target spacecraft that was launched on board an 1165M Kosmos-2 two-stage launch vehicle from the Plesetsk cosmodrome in Soviet Russia on 9 July 1976.

The mission had an epoch start time/date of 21:07:00 UTC on 9 July 1976 (conflicting NASA data states an epoch start time/date of 20:00:00 UTC on 8 July). The mission orbital parameters included a periapsis of 984 km, apoapsis of 2102 km, period 117 minutes, inclination 65.9° and an eccentricity of 0.07053. There is no reliable information regarding orbital decay date.

Cosmos 843: Cosmos 843 was a second generation IS interceptor spacecraft that was launched on board a Cyclone-2 two-stage launch vehicle from the Baikonur cosmodrome at Tyuratam in Soviet Kazakhstan on 21 July 1976. The mission was to intercept the Cosmos 839 ASAT target spacecraft that had been launched on 9 July.

The mission had an epoch start time/date of 15:21:00 UTC on 21 July 1976. The mission orbital parameters included a periapsis of 149 km, apoapsis of 360 km, period 89.4 minutes, inclination 65.1° and an eccentricity of 0.01589. The above orbital parameters are not consistent with an interception profile on Cosmos 839, suggesting that the mission encountered problems of one sort or another. The spacecraft was de-orbited later on 21 July 1976.

Cosmos 880: Cosmos 880 was a Tjulpan adjustment and calibration ASAT target spacecraft that was launched on board an 1165M Kosmos-2 two-stage launch vehicle from the Plesetsk cosmodrome in Soviet Russia on 9 December 1976. The spacecraft mission was to act as a target for the Cosmos 886 IS interceptor that would be launched a few weeks after Cosmos 880.

The mission had an epoch start time/date of 20:09:00 UTC on 9 December 1976. The mission orbital parameters included a periapsis of 526 km, apoapsis of 624 km, period 96.4 minutes, inclination 66° and an eccentricity of 0.00704. The spacecraft orbit did not decay until 31 August 1991.

Cosmos 886: Cosmos 886 was a second generation IS interceptor spacecraft that was launched on board a Cyclone-2 two-stage launch vehicle from the Baikonur cosmodrome at Tyuratam in Soviet Kazakhstan on 27 December 1976. The mission was to intercept the Cosmos 880 ASAT target spacecraft that had been launched on 9 December that year.

The mission had an epoch start time/date of 19:00:00 UTC on 27 December 1976. The mission orbital parameters included a periapsis of 581 km, apoapsis of 2328 km, period 115 minutes, inclination 66° and an eccentricity of 0.11145. Following a successful interception of Cosmos 880, Cosmos 886 was de-orbited later on 27 December 1976.

Cosmos 909: Cosmos 909 was a Tjulpan adjustment and calibration ASAT target spacecraft that was launched on board an 1165M Kosmos-2 two-stage launch vehicle from the Plesetsk cosmodrome in Soviet Russia on 19 May 1977. The spacecraft mission was to act as a target for the Cosmos 910 IS interceptor spacecraft that would be launched several days later.

The mission had an epoch start time/date of 16:38:00 UTC on 19 May 1977 (conflicting NASA documentation states an epoch start time/date of 20:00:00 UTC on 18 May 1977). The mission orbital parameters included a periapsis of 991 km, apoapsis of 2112 km, period 117 minutes, inclination 65.9° and an eccentricity of 0.07064. There is no reliable data on orbital decay date.

Cosmos 910: Cosmos 910 was a second generation IS interceptor spacecraft that was launched on board a Cyclone-2 two-stage launch vehicle from the Baikonur cosmodrome at Tyuratam in Soviet Kazakhstan on 23 May 1977. The mission was to intercept the Cosmos 909 ASAT target spacecraft that had been launched several days earlier.

The mission had an epoch start time/date of 12:38:00 UTC on 23 May 1977. The mission orbital parameters included a periapsis of 149 km, apoapsis of 506 km, period 91 minutes, inclination 65.1° and an eccentricity of 0.0266. The best available data would suggest that the spacecraft failed to intercept Cosmos 909 as the orbital parameters noted above are inconsistent with an orbital interception profile for Cosmos 909. Cosmos 910 was de-orbited later on 23 May 1977.

Cosmos 918: Cosmos 918 was a second generation IS interceptor spacecraft that was launched on board a Cyclone-2 two-stage launch vehicle from the Baikonur cosmodrome at Tyuratam in Soviet Kazakhstan on 17 June 1977. It is assumed that the mission objective was to intercept the Cosmos 909 ASAT target spacecraft that was launched the previous month. This mission, like the Cosmos 910 mission before it, appears to have failed to intercept the target as it did not achieve the required orbital parameters.

The mission had an epoch start time/date of 07:26:00 UTC on 17 June 1977. The mission orbital parameters included a periapsis of 131 km, apoapsis of 265 km, period 88.4 minutes, inclination 65.1° and an eccentricity of 0.01018. The orbital parameters noted above are inconsistent with an orbital interception profile for Cosmos 909 supporting the conclusion that the mission failed. Cosmos 918 was de-orbited on 18 June 1977.

Cosmos 959: Cosmos 959 was a Tjulpan adjustment and calibration ASAT target spacecraft that was launched on board an 1165M Kosmos-2 two-stage launch vehicle from the Plesetsk cosmodrome in Soviet Russia on 21 October 1977. The spacecraft mission was to act as a target for the Cosmos 961 IS interceptor spacecraft that would be launched several days later.

The mission had an epoch start time/date of 10:04:00 UTC on 21 October 1977 (conflicting NASA documentation states an epoch start date/time of 20:00:00 UTC on 20 October 1977). The mission orbital parameters included a periapsis of 146 km, apoapsis of 850 km, period 94.57 minutes, inclination 65.84° and an eccentricity of 0.05116. The Cosmos 959 spacecraft orbit decayed on 30 November 1977.

Cosmos 961: Cosmos 961 was a second generation IS interceptor spacecraft launched on board a Cyclone-2 two-stage launch vehicle from the Baikonur cosmodrome at Tyuratam in Soviet Kazakhstan on 26 October 1977. The mission was to intercept the Cosmos 959 target that had been launched several days prior.

The mission had an epoch start time/date of 05:16:00 UTC on 26 October 1977 (conflicting NASA documentation states an epoch start of 20:00:00 UTC on 25 October 1977). The mission orbital parameters included a periapsis of 125 km, apoapsis of 302 km, period 88.5 minutes, inclination 66° and an eccentricity of 0.01342. The spacecraft was de-orbited 26 October 1977.

Cosmos 967: Cosmos 967 was a Tjulpan adjustment and calibration ASAT target spacecraft that was launched on board an 1165M Kosmos-2 two-stage launch vehicle from the Plesetsk cosmodrome in Soviet Russia on 13 December 1977. The spacecraft mission was to act as a target for the Cosmos 970 IS interceptor spacecraft that would be launched just over one week later.

The Cosmos 967 mission had an epoch start time/date of 15:50:00 UTC on 13 December 1977 (conflicting NASA documentation states an epoch start of 19:00:00 UTC on 12 December 1977). The mission orbital parameters included a periapsis of 973 km, apoapsis of 1013 km, period 105 minutes, inclination 66° and an eccentricity of 0.00271. There is no reliable data available for orbital decay date.

Cosmos 970: Cosmos 970 was a second generation IS interceptor spacecraft that was launched on board a Cyclone-2 two-stage launch vehicle from the Baikonur cosmodrome at Tyuratam in Soviet Kazakhstan on 21 December 1977. The mission was to intercept the Cosmos 967 ASAT target spacecraft launched just over one week before.

The mission had an epoch start time/date of 10:33:00 UTC on 21 December 1977 (conflicting NASA documentation states an epoch start time/date of 19:00:00 UTC on 20 December 1977). The mission orbital parameters included a periapsis of 144 km, apoapsis of 861 km, period 94.67 minutes, inclination 65.166° and an eccentricity of 0.05207. At the end of the mission Cosmos 970 conducted an engine burn series and was de-orbited.

Cosmos 1009: Cosmos 1009 was a second generation IS interceptor spacecraft that was launched on board a Cyclone-2 two-stage launch vehicle from the Baikonur cosmodrome at Tyuratam in Soviet Kazakhstan on 19 May 1978. NASA documentation states that Cosmos 1009 intercepted Cosmos 970, but this must be erroneous as Cosmos 970 was de-orbited on 21 December 1977. The orbital parameters for Cosmos 1009 were more consistent with an interception of Cosmos 967. In the absence of other reliable data it seems reasonable to conclude that it was indeed the Cosmos 967 ASAT target that was intercepted by Cosmos 1009.

The mission had an epoch start time/date of 00:28:00 UTC on 19 May 1978 (conflicting NASA documentation states an epoch start time/date of 20:00:00 UTC on 18 May 1978). The mission orbital parameters included a periapsis of 971 km, apoapsis of 1378 km, period 109 minutes, inclination 66° and an eccentricity of 0.02693. The spacecraft was de-orbited on 19 May 1978.

The Soviet Union introduced a unilateral moratorium on ASAT testing, which came into effect at the conclusion of the Cosmos 1009 mission on 19 May 1978. This was a gesture aimed at improving the atmosphere of inter-superpower relations in advance of future arms control/limitation talks with the United States. By the time of the Soviet-American talks in Vienna on 23 April 1979, in effect the third round of Soviet-American space-warfare talks, the Soviets had not conducted ASAT testing for close to one year.

The main goal for the American side was to try and get a one year ban on such testing; in effect delay further Soviet testing until the United States was better positioned to respond with its own ASAT programs, none of which were, at that time, near the flight test phase. The most promising American program at that time was a Vought development nicknamed the 'Flying Tomato Can' in Pentagon circles. This system, which the Americans considered would be superior in operational capability to the deployed Soviet Co-Orbital Counter Space Defence Satellite Fighter complex as it was being developed a generation later, was expected to home in on a satellites heat signature and destroy it in a collision. Development and flight testing of the 'Flying Tomato Can' program had been approved by President Ford. However when President Carter took office there was a bit of foot shuffling as he contemplated continuing development, but not proceeding to flight tests status.

However, in spring 1977, President Carter continued with endorsement of the program all the way to flight test status.

In 1979, the best available estimates were for initial flight testing to commence sometime in 1980, the Americans hoping for a unilateral Soviet ASAT test flight ban until then. At this time there seemed, despite the plethora of observational data available, either a genuine misconception of the altitude capabilities of the Soviet ASAT system, or a deliberate intelligence misinformation program. The evidence for this comes from advice from so called experts that stated that the Soviet system had a capability against low-orbit satellites operating at altitudes up to around 193 km, when, in actuality, the system had been tested up to orbital altitudes considerably in excess of this.

With intelligence assessments that the United States was pushing ahead with its own ASAT programs, the Soviets ended their unilateral moratorium on ASAT testing. Plans for a new phase of testing were dusted-off and prepared for implementation, which commenced with the launch of Cosmos 1171 on 3 April 1980.

Cosmos 1171: Cosmos 1171 was a Tjulpan adjustment and calibration ASAT target spacecraft that was launched on board an 1165M Kosmos-2 two-stage launch vehicle from the Plesetsk cosmodrome in Soviet Russia on 3 April 1980. The spacecraft mission was to act as a target for the Cosmos 1174 IS interceptor spacecraft that would be launched a few weeks later.

The mission had an epoch start time/date of 07:40:00 UTC on 3 April 1980 (conflicting NASA documentation states an epoch start time/date of 19:00:00 UTC on 2 April 1980). The mission orbital parameters included a periapsis of 976 km, apoapsis of 1017 km, period 105 minutes, inclination 65.8° and an eccentricity of 0.00278. There is no reliable information for orbital decay date.

Cosmos 1174: Cosmos 1174 was a second generation IS interceptor spacecraft that was launched on board a Cyclone-2 two-stage launch vehicle from the Baikonur cosmodrome at Tyuratam in Soviet Kazakhstan on 18 April 1980. This was the first operational test of an IS interceptor designed to test the spacecraft functionality following a lengthy period of storage.

Cosmos 1174 was put through a series of maneuvering phases over the next few days, no passes sufficiently close enough to Cosmos 1171 to be considered a successful interception being attained. Cosmos 1174 appears to have been remote exploded on 20 April 1980, although it is unclear if this was a mission intent or perhaps as a result of problems being encountered on a re-entry attempt.

The mission had an epoch start time/date of 00:31:00 UTC on 18 April 1980 (conflicting NASA documentation states an epoch start time/date of 19:00:00 UTC on 17 April 1980). The mission orbital parameters included a periapsis of 387 km, apoapsis of 1035 km, period 98.6 minutes, inclination 65.8° and an eccentricity of 0.04567.

Cosmos 1241: Cosmos 1241 was a Tjulpan adjustment and calibration ASAT target spacecraft that was launched on board an 1165M Kosmos-2 two-stage launch vehicle from the Plesetsk cosmodrome in Soviet Russia on 21 January 1981. The spacecraft mission was to act as a target for the Cosmos 1243 interceptor (launched on 2 February 1981) and later the Cosmos 1258 IS interceptor (launched on 14 March 1981).

The mission had an epoch start time/date of 08:24:00 UTC on 21 January 1981 (conflicting NASA documentation states an epoch start time/date of 19:00:00 UTC on 20 January 1981). The mission orbital parameters included a periapsis of 1000 km, apoapsis of 1000 km, period 105 minutes, inclination 65.8° and an eccentricity of 0.

Cosmos 1243: Cosmos 1243 was a second generation IS interceptor spacecraft launched on board a Cyclone-2 two-stage launch vehicle from the Baikonur cosmodrome at Tyuratam in Soviet Kazakhstan on 2 February 1981. This mission was designed to test the spacecraft functionality following a lengthy period of storage. The target for the interceptor was the Cosmos 1241 ASAT target spacecraft that had been launched on 21 January 1981. During the interception, Cosmos 1243 closed to within 50 m of the Cosmos 1241 target, but the fragmentation warhead designed to destroy the target apparently failed to detonate. Following the test phase the Cosmos 1243 interceptor was de-orbited and burned up in the atmosphere on 2 February 1981.

The mission had an epoch start time/date of 02:24:00 UTC on 2 February 1981 (conflicting NASA documentation states an epoch start time/date of 19:00:00 UTC on 1 February 1981). The mission orbital parameters included a periapsis of 316 km, apoapsis of 1026 km, period 98 minutes, inclination 66° and an eccentricity of 0.05033.

Cosmos 1258: Cosmos 1258 was a second generation IS interceptor spacecraft that was launched on board a Cyclone-2 two-stage launch vehicle from the Baikonur cosmodrome at Tyuratam in Soviet Kazakhstan on 14 March 1981. The mission employed the Cosmos 1241 ASAT target that was still in orbit, but available data suggests that no successful interception took place.

The mission had an epoch start time/date of 17:02:00 UTC on 14 March 1981 (conflicting NASA documentation states an epoch start time/date of 19:00:00 UTC on 13 March 1981). The mission orbital parameters included a periapsis of 322 km, apoapsis of 1032 km, period 98 minutes, inclination 65.8° and an eccentricity of 0.05028. Following the test manoeuvres Cosmos 1258 was maneuvered to de-orbit and burned up in the atmosphere on 15 March 1981.

Cosmos 1375: Cosmos 1375 was a Tjulpan adjustment and calibration ASAT target spacecraft that was launched on board an 1165M Kosmos-2 two-stage launch vehicle from the Plesetsk cosmodrome in Soviet Russia on 6 June 1982. The spacecraft mission was to act as a target for the Cosmos 1379 IS interceptor that would be launched on 18 June 1982.

The mission had an epoch start time/date of 17:02:00 UTC on 6 June 1982 (conflicting NASA documentation states an epoch start time/date of 20:00:00 UTC on 5 June 1982). The mission orbital parameters included a periapsis of 990 km, apoapsis of 1021 km, period 105 minutes, inclination 65.9° and an eccentricity of 0.0021.

Cosmos 1379: Cosmos 1379 was a second generation IS interceptor spacecraft that was launched on board a Cyclone-2 two-stage launch vehicle from the Baikonur cosmodrome at Tyuratam in Soviet Kazakhstan on 18 June 1982. The mission was designed to target and intercept the Cosmos 1375 ASAT target spacecraft launched almost two weeks before.

The mission had an epoch start time/date of 11:04:00 UTC on 18 June 1982 (conflicting NASA documentation states an epoch start time/date of 20:00:00 UTC on 17 June 1982). The mission orbital parameters included a periapsis of 552 km, apoapsis of 1027 km, period 100.3 minutes, inclination 65.8° and an eccentricity of 0.03311. Following a successful interception of Cosmos 1375, Cosmos 1379 was de-orbited later on 18 June, bringing to an end the flight testing of the IS interceptor spacecraft of the Co-Orbital Counter Space Defence complex.

The Cosmos 1379-Cosmos-1375 mission was the final live test demonstration of the Soviet and later Russian Co-Orbital Satellite Fighter capability. However, the system remained on alert status through the remainder of the Cold War, which it outlived, being retired from the space defence forces of the Russian Federation that gad taken on the mantle of the space defence programs of the Soviet Union following the dissolution of the USSR (Union of Soviet Socialist Republics) on 25 December 1991. Following some two decades of operational service on alert status, the Counter-Space Defence complex was stood down in 1993, the harsh economic conditions that prevailed in the Russian Federation in the immediate post-Cold War years leading to the cancellation of the Counter-Space Defence complex planned successor. However, in 2016/2017, there have been a number of references to the establishment of a new space defence system, although whether this will be in the form of a terrestrial based direct assent missile system or a 21st Century incarnation of the Cold War era Counter-Space Defence interceptor, remains unclear in early 2017.

GLOSSARY

ABM	Anti-Ballistic Missile
Apoapsis	The farthest point reached by an orbiting object from the central body being orbited (also known as apogee)
ASAT	Anti-Satellite
CIA	Central Intelligence Agency
Cosmos	Designation used for many Soviet and later Russian Federation space vehicles that operated in Earth orbit
DIA	Defence Intelligence Agency
Eccentricity	Measurement of the extent to which an orbiting object departs from an ellipse from a relative circle. This measurement is calculated as 1 full half of the distance of the two foci (the two fixed points of the ellipse) divided by the length of the semi-major axis. The calculated number is dimensionless, carrying values between 0 and 1.
EDB	Experimental Design Bureau
Epoch	Effectively the start of the timeline of a particular period of time or event
FOBS	Fractional Orbit Bombardment System
ICBM	Intercontinental Ballistic Missile
II	Roman numeral number 2
Inclination	In regards to an orbit, it refers to the angle created between the plane of the orbit and the ecliptic plane
IP	Istrebitel Sputnik
IS	Istrebitel Sputnik
JSC	Joint Stock Company
kg	Kilogram
kg/cm^2	Kilogram per centimeter square
km	Kilometre
LEO	Low Earth Orbit
LRBM	Long Range Ballistic Missile
m	Metre
MAD	Mutually Assured Destruction
mm	Millimetre
MOBS	Multiple Orbit Bombardment System
MODRF	Ministry of Defence of the Russian Federation
MRBM	Medium Range Ballistic Missile
NASA	National Aeronautics and Space Administration
NATO	North Atlantic Treaty Organisation
nm	Nautical Mile
NTO	Nitrogen Tetroxide
OJSC	Open Joint Stock Company
OKB	Design Bureau - Experimental Design Bureau
OKB-1	Experimental Design Bureau-1 (now S.P. Korolev

Rocket and Space Corporation, Energia)

OKB-52   Experimental Design Bureau-52 (now JSC MIC Mashinostroyenia (Joint Stock Company Military Industrial Corporation Scientific and Production Machine Building Association)

OKB-586   Experimental Design Burea-586 (now Yuzhnoye State Design Office)

OKO   Ground Control Centre

Periapsis   The closest approach point reached by an orbiting object from the central body being orbited

Period   The orbital period refers to the time taken for an object to complete one full revolution around its orbit. The time can be measured in seconds, minutes, days or years

R   Rocket

SRBM   Short Range Ballistic Missile

TMBDB   Turaevo Machine Building Design Bureau

tnf   Ton force

UDMH   Unsymmetrical Dimethylhydrazine

UK   United Kingdom

US   United States

UTC   Coordinated Universal Time

Warsaw Pact   A formal treaty of friendship, co-operation and mutual assistance signed between the Socialist Republics of the USSR and 7 Soviet Orbit satellite states in Eastern Europe. This treaty, which took effect from 14 May 1955, was designed to counter the growing NATO alliance opposed to the Eastern Block

X   Experimental

°   Degree(s)

~   Approximately equal to (can also be used to mean asymptotically equal)

ABOUT THE AUTHOR

Hugh, a historian and author with an extensive background in astro/geophysics and studies/research in the wider scientific, aeronautic, astronautic and nautical technical and historical fields, has published in excess of sixty books; non-fiction and fiction, writing under his given name as well as utilising several pseudonyms. He has also written for several international magazines, whilst his work has been used as reference for many other projects ranging from the aviation industry, international news corporations and film media to encyclopaedias, museum exhibits and the computer gaming industry. Hugh is a member of the institute of Physics, a member of the British Geophysics Association and is an elected Fellow of the Royal Astronomical Society. He currently resides in his native Scotland.

Other titles by the author include
Sukhoi T-50/PAK FA - Russia's 5th Generation 'Stealth' Fighter
Sukhoi Su-35S 'Flanker' E - Russia's 4++ Generation Super-Manoeuvrability Fighter
Sukhoi Su-34 'Fullback'
Sukhoi Su-30MKK/MK2/M2 - Russo Kitashiy Striker from Amur
MiG-35/D 'Fulcrum' F – Towards the Fifth Generation
Air War over Syria, Tu-160, Tu-95MS & Tu-22M3 - Cruise Missile and Bombing Strikes on Syria, November 2015-February 2016
Sukhoi Su-27SM(3)/SKM
Iskander - Mobile Tactical Aero-Ballistic/Cruise Missile Complex
Orbital/Fractional Orbit Bombardment System - The Soviet Globalnaya Raketa
Russian Non-Nuclear Attack Submarines – Project 877/877E/877EKM/Project 636/636.3 & Project 677/Amur 1650/950/S-1000
Russian/Soviet Aircraft Carrier & Carrier Aviation Design & Evolution Volume 1 - Seaplane Carriers, Project 71/72, Graf Zeppelin, Project 1123 ASW Cruiser & Project 1143-1143.4 Heavy Aircraft Carrying Cruiser
Light Battle Cruisers and the Second Battle of Heligoland Bight
British Battlecruisers of World War 1 - Operational Log, July 1914-June 1915
Eurofighter Typhoon - Storm over Europe
Tornado F.2/F.3 Air Defence Variant
Air to Air Missile Directory
North American F-108 Rapier - Mach 3 Interceptor
Convair YB-60 - Fort Worth Overcast
Boeing X-36 Tailless Agility Flight Research Aircraft
X-32 - The Boeing Joint Strike Fighter
X-35 - Progenitor to the F-35 Lightning II
X-45 Uninhabited Combat Air Vehicle
Into The Cauldron - The Lancaster MK.I Daylight Raid on Augsburg
Hurricane IIB Combat Log - 151 Wing RAF, North Russia 1941
RAF Meteor Jet Fighters in World War II, an Operational Log
Typhoon IA/B Combat Log - Operation Jubilee, August 1942
Defiant MK.I Combat Log - Fighter Command, May-September 1940
Blenheim MK.IF Combat Log - Fighter Command Day Fighter Sweeps/Night Interceptions, September 1939 - June 1940
Tomahawk I/II Combat Log - European Theatre, 1941-42
Fortress MK.I Combat Log - Bomber Command High Altitude Bombing Operations, July-September 1941
XF-92 - Convairs Arrow